Land Surface Evaluation for Engineering Practice

Geological Society Special Publications
Series Editor
J. S. GRIFFITHS

Special Publication reviewing procedures

The Society makes every effort to ensure that the scientific and production quality of its books matches that of its journals. Since 1997, all book proposals have been refereed by specialist reviewers as well as by the Society's Publications Committee. If the referees identify weaknesses in the proposal, these must be addressed before the proposal is accepted.

Once the book is accepted, the Society has a team of series editors (listed above) who ensure that the volume editors follow strict guidelines on refereeing and quality control. We insist that individual papers can only be accepted after satisfactory review by two independent referees. The questions on the review forms are similar to those for *Quarterly Journal of Engineering Geology and Hydrogeology*. The referees' forms and comments must be available to the Society's series editors on request.

Although many of the books result from meetings, the editors are expected to commission papers that were not presented at the meeting to ensure that the book provides a balanced coverage of the subject. Being accepted for presentation at the meeting does not guarantee inclusion in the book.

Geological Society Special Publications are included in the ISI Science Citation Index, but they do not have an impact factor, the latter being applicable only to journals.

More information about submitting a proposal and producing a Special Publication can be found on the Society's web site: www.geolsoc.org.uk.

It is recommended that reference to all or part of this book should be made in one of the following ways:

GRIFFITHS, J. S. (ed.) 2001. *Land Surface Evaluation for Engineering Practice*. Geological Society, London, Engineering Geology Special Publications, **18**.

FOOKES, P. G. & SHILSTON, D. T. 2001. Building the geological model: case study of a rock tunnel in SW England. *In*: GRIFFITHS, J. S. (ed.) *Land Surface Evaluation for Engineering Practice*. Geological Society, London, Engineering Geology Special Publications, **18**, 123–128.

Geological Society Engineering Geology Special Publication No. 18

Land Surface Evaluation for Engineering Practice

EDITED BY

J. S. GRIFFITHS

Department of Geological Sciences
University of Plymouth
Drake Circus
Plymouth
Devon PL4 8AA

2001
Published by
The Geological Society
London

THE GEOLOGICAL SOCIETY

The Geological Society of London was founded in 1807 and is the oldest geological society in the world. It received its Royal Charter in 1825 for the purpose of 'investigating the mineral structure of the Earth' and is now Britain's national society for geology.

Both a learned society and a professional body, the Geological Society is recognized by the Department of Trade and Industry (DTI) as the chartering authority for geoscience, able to award Chartered Geologist status upon appropriately qualified Fellows. The Society has a membership of 9099, of whom about 1500 live outside the UK.

Fellowship of the Society is open to persons holding a recognized honours degree in geology or cognate subject, or not less than six years' relevant experience in geology or a cognate subject. A Fellow with a minimum of five years' relevant postgraduate experience in the practice of geology may apply for chartered status. Successful applicants are entitled to use the designatory postnominal CGeol (Chartered Geologist). Fellows of the Society may use the letters FGS. Other grades of membership are available to members not yet qualifying for Fellowship.

The Society has its own Publishing House based in Bath, UK. It produces the Society's international journals, books and maps, and is the European distributor for publications of the American Association of Petroleum Geologists (AAPG), the Society for Sedimentary Geology (SEPM) and the Geological Society of America (GSA). Members of the Society can buy books at considerable discounts. The Publishing House has an online bookshop (*http://bookshop.geolsoc.org.uk*).

Further information on Society membership may be obtained from the Membership Services Manager, The Geological Society, Burlington House, Piccadilly, London W1V 0JU (E-mail: *enquiries@geolsoc.org.uk*; tel: +44 (0) 207 434 9944).

The Society's Web Site can be found at *http://www.geolsoc.org.uk/*. The Society is a Registered Charity, number 210161.

Published by The Geological Society from:
The Geological Society Publishing House
Unit 7 Brassmill Enterprise Centre
Brassmill Lane
Bath BA1 3JN, UK
(*Orders*: Tel. +44 (0)1225 445046
        Fax +44 (0)1225 442836)
Online bookshop: *http://bookshop.geolsoc.org.uk*

British Library Cataloguing in Publication Data
A catalogue record for this book is available from the British Library.

ISBN 1–86239–084–3
ISSN 0267–9914

Typeset by Aarontype Ltd, Bristol BS5 0HE, UK.

Printed by The Alden Press, Osney Mead, Oxford, UK.

Distributors

USA
AAPG Bookstore
PO Box 979
Tulsa
OK 74101-0979
USA
(*Orders*: Tel. +1 918 584-2555
        Fax +1 918 560-2652
        E-mail *bookstore@aapg.org*

Australia
Australian Mineral Foundation Bookshop
63 Conyngham Street
Glenside
South Australia 5065
Australia
(*Orders*: Tel. +61 88 379-0444
        Fax +61 88 379-4634
        E-mail *bookshop@amf.com.au*

India
Affiliated East-West Press PVT Ltd
G-1/16 Ansari Road, Daryaganj
New Delhi 110 002
India
(*Orders*: Tel. +91 11 327-9113
        Fax +91 11 326-0538
        E-mail *affiliat@nda.vsnl.net.in*

Japan
Kanda Book Trading Co.
Cityhouse Tama 204
Tsurumaki 1-3-10
Tauama-shi
Tokyo 206-0034
Japan
(*Orders*: Tel. +81 (0)423 57-7650
        Fax +81 (0)423 57-7651

Contents

Preface
The Second Working Party on Land Surface Evaluation for Engineering Practice

The original Working Party on Land Surface Evaluation for Engineering Practice, under the chairmanship of **Mr R. J. G. Edwards**, reported in 1982. Their report was presented in the *Quarterly Journal of Engineering Geology*, Volume 15, pages 265–316, and was based primarily on work that had been carried out during the previous decade. Since the original report was published there have been considerable advances in the subject. Therefore, the Committee of the Engineering Group of the Geological Society set up the Second Working Party in January 1997. The membership of the Second Working Party comprised the following.

Dr J. S. Griffiths: Chairman: Head of the Department of Geological Sciences at the University of Plymouth and member of Plymouth Environmental Research Group; Committee Member of the Engineering Group of the Geological Society (1997–2003).

Mr R. J. G. Edwards: Chairman of the First Working Party; Consultant Engineering Geologist; Director of Earth Science Partnership.

Professor D. Brunsden: Emeritus Professor in Physical Geography at Kings College, London; Member of the First Working Party; First President of the International Association of Geomorphologists; and the Fifth Glossop Lecturer (2001).

Mr J. H. Charman: Consultant Engineering Geologist; Chairman of the Engineering Group 1996–98; Member of the Council of the Geological Society (1998–2000).

Dr P. Nathanail: Senior Lecturer at Nottingham University; Chairman of the Environment Group of the Geological Society (1997–99).

Mr W. Rankin: Divisional Director of Foundations and Geotechnics at Mott MacDonald; Chairman of the Association of Geotechnical Specialists (1996–98); Chief Executive of AGS (1998–2000); Chairman of Ground Forum (2000–2001).

Mr P. Phipps: Senior Engineer at Mott MacDonald; First Glossop Award Winner (1997); Committee Member of the Engineering Group (1997–2000).

During the early meetings of the Working Party it became apparent that the report would have to take a different form from the normal Engineering Group Working Party publications. It was decided that the range and breadth of the subject would require input from a large number of specialists if it was to provide a realistic view of the state of the art. The result was that practitioners in land surface evaluation known to the members of the Working Party were invited to contribute short papers to an edited volume.

In the compilation of the final edited volume, substantial assistance was provided by **Dr Gareth J. Hearn** of Scott Wilson Kirkpatrick, and **Mr E. Mark Lee** of the University of Newcastle, who, along with **Professor Denys Brunsden**, acted as sub-editors.

All members of the Working Party and the sub-editors gave freely of their time in the preparation of this report. The backing of employers in giving professionals the space and time on this, and similar working parties, is too often taken for granted. As Chairman of the Working Party I wish to record the debt of gratitude that I owe to the many individuals, companies and university departments who have provided such support. In addition, I wish particularly to thank **Mott MacDonald and Scott Wilson Kirkpatrick** for providing financial backing towards the cost of printing the final report, thus allowing us to use colour in some of the figures.

Dr James S Griffiths
Chairman
University of Plymouth
October 2000

Section 1

Introduction

The development of land surface evaluation for engineering practice

J. S. Griffiths[1] & R. J. G. Edwards[2]

[1] Department of Geological Sciences, University of Plymouth, Devon, UK
[2] Earth Sciences Partnership, Leatherhead, Surrey, UK

Definition

The First Working Party Report (Anon. 1982) defined 'Land surface evaluation' for engineering practice as: '*The evaluation and interpretation of land surface features and recorded surface data using one or a combination of the ground mapping, interpretation, classification and visual remote sensing techniques outlined in this report*'. The techniques outlined in the report were land classification, remote sensing and geomorphological mapping. The expression 'land surface evaluation' was adopted in preference to 'terrain evaluation' or 'terrain classification' because the varied uses of the terms had created confusion and led to misunderstandings. For this same reason in this Second Working Party Report the expression 'land surface evaluation' has been maintained. Developments since the 1982 Report was published require revision to the definition of 'land surface evaluation'. The definition proposed by the Second Working Party is: '*The evaluation and interpretation of land surface and near surface features using techniques that do not involve ground exploration by excavation or geophysics*'. This rather broader definition allows land surface evaluation to be seen in its most common context as the process of data compilation, interpretation and conceptual ground modelling prior to undertaking engineering ground or site investigation work. It therefore specifically includes the integration of all existing ground information (desk study) whether it is surface or subsurface information.

The first objective of a land surface evaluation study is to acquire the most comprehensive conceptual ground model that can be generated in order to maximize the value and justify the cost of subsurface investigation fieldwork and allied laboratory testing. The second objective is to minimize the engineering geological unknowns that currently generate disproportionate unforeseen contractual costs and not infrequently jeopardize the in-service facility or its design life.

Role of land surface evaluation

The 1982 Report stated that '*Land surface evaluation should be regarded and used as a normal method of investigation for ground engineering*'. The 1982 Report identified the method as an integral part of the site investigation process typically undertaken at the reconnaissance and feasibility study stages of a project. The techniques were also seen as a specific method of data collection and compilation for selected aspects of investigation or monitoring at the design, construction and post-construction stages. The Second Working Party fully endorses this view but, on the basis of identified practice, has identified that land surface evaluation techniques also have a clear role to play in general planning of engineering developments.

Land surface evaluation techniques recommended by the 1982 Report

The main land surface evaluation methods recommended in the 1982 Report were geomorphological ground mapping and aerial photographic interpretation based on a framework of land classification. These recommendations arose out of the experience gained on projects undertaken during the previous four decades. Of particular importance was the initial development of aerial photographic interpretative procedures (Belcher 1948) and the subsequent geomorphologically based developments in land classification or terrain systems (Mitchell 1973). In engineering and resource survey projects the use of land classification based on terrain systems was pioneered by the Transportation Road Research Laboratory (TRRL) (Dowling 1968; Dowling & Bevan 1969). Other important contributions to the subject included: the development of land unit maps for data banks in South Africa (Brink & Partridge 1967; Brink *et al.* 1968; National Institute for Road Research 1971); the PUCE terrain classification system in Australia (Aitchison & Grant 1967, 1968); and the Oxford–MEXE–Cambridge System of Classification (Beckett & Webster 1969; MEXE 1969; Perrin & Mitchell 1969). The MEXE (Military Experimental Establishment) system was developed for military use and adopted by the TRRL for the location of highways in the humid tropics (Lawrence 1978). It is interesting to note that, whilst not

From: GRIFFITHS, J. S. (ed.) *Land Surface Evaluation for Engineering Practice*. Geological Society, London, Engineering Geology Special Publications, **18**, 3–9. 0267-9914/01/$15.00 © The Geological Society of London 2001.

set out in specific detail, the value of these techniques was referred to in the original 1950 Civil Engineering Code of Practice No. 1 updated for site investigations as CP2001 (British Standards Institution 1957).

Most of the initial land classification studies were undertaken at a reconnaissance level using relatively small-scale mapping (less than 1:25 000). During the 1970s larger scale mapping specifically for engineering projects was introduced to facilitate conceptual ground modelling for planning site investigation work and it also became established as an important tool in the interpretation and presentation of site investigation output. A strong impetus to the wider use of mapping techniques, particularly in engineering practice, was provided by the Engineering Group Working Party Report on maps and plans (Anon. 1972). Although not specifically addressed, many of the recommendations for good practice that were included in the Working Party Report involved the application of land surface evaluation techniques. During the 1970s a range of mapping techniques were developed for engineering practice using both field mapping and aerial photographic interpretation. These developments included engineering geological mapping (Dearman & Fookes 1974), geotechnical mapping (Clark & Johnson 1975), engineering geomorphological mapping (Brunsden et al. 1975; Doornkamp et al. 1979), and refinements in the use of aerial photographic interpretation for the analysis of engineering geological problems (Edwards 1968; Verstappen & Van Zuidan 1968).

The First Working Party Report, therefore, was published at a time when there was cause for cautious optimism that the value and range of potential uses of land surface evaluation in engineering practice as an integral part of site investigations was being recognized. However, in 1981, BS5930 was published by the British Standards Institution (1981) and this Code of Practice has formed the basis of site investigation for civil engineering in the UK for the 18 years up to the publication of the revised code in 1999 (British Standards Institution 1999). Although the importance of desk studies and the need for rational site investigation planning was identified, BS5930: 1981 clearly placed the emphasis on ground investigation using exploratory holes, geophysics, and field and laboratory testing. The first edition of BS5930 drew attention to the use of aerial photography, and geomorphological and engineering geological mapping, but it made no reference to the importance of integrating this information using land surface evaluation techniques to generate conceptual or predictive ground models. This may partly explain why the expansion in the use of land surface evaluation techniques envisaged in the 1970s, and recommended in the 1982 Report, appears to have been rather limited, as indicated by the published case studies.

Developments in the UK since 1982

Throughout the 1980s and 1990s a significant number of engineering case study applications of the recommended techniques of land surface evaluation were reported in the literature (e.g. Jones et al. 1983; Fookes et al. 1985; Griffiths & Marsh 1986; Charman & Griffiths 1993). However, it is in the Applied Earth Science Mapping programme carried out by various companies for the Department of the Environment (DoE) (e.g. Doornkamp 1988; Forster et al. 1987; Wallace Evans Ltd 1994) that the most widespread use of land surface evaluation techniques can be found. These DoE studies demonstrated that the techniques could be directly applied in planning development and for engineering and environmental studies (Brook & Marker 1987; Dearman 1987; DoE 1991). The range of scale adopted for these studies varied but commonly involved compilation of a suite of 'earth science' maps at a scale of 1:25 000. One of the few larger scale studies (maps at 1:2500 scale) was carried out for the assessment of landslide risk on the Isle of Wight at Ventnor (Lee et al. 1991). This programme of applied earth science mapping has continued through the 1990s, mainly carried out by the British Geological Survey, and there are approximately 50 such studies available for areas of the UK (Smith & Ellison 1999). The DoE programme of earth science mapping and the various detailed studies of ground instability (e.g. GSL 1987) formed the basis for the publication of PPG14 and PPG14 Annex 1 (DoE 1990, 1995). These notes represent the Government's planning guidance concerned with the granting of planning permission for construction on unstable ground.

Whilst land surface evaluation techniques clearly made an important contribution to planning studies during the 1980s and 1990s, the published records of their use in engineering practice has been disappointing. The development of engineering geomorphology in the UK, seen in the 1982 Working Party Report as a central component of land surface evaluation, was discussed in Griffiths & Hearn (1990). This study indicated that during the eight years since the publication of the First Working Party Report, the use of geomorphology in engineering practice had not been particularly widespread. The causes of this were complex. In part it was due to a lack of appreciation by engineers of how important and cost effective ground modelling is in the development of investigatory works for engineering projects. This was exacerbated by a continuing failure amongst clients to recognize that money spent wisely in the initial phases of project development pays massive dividends in overall site investigation, design, construction and maintenance costs. In part it was also due to a lack of opportunity for such specialists to prove the value of the technique because of failures in communication between geomorphologists with a geographic background, and engineers. In addition, such specialists frequently did not have the training or understanding of either geology or engineering to be able to generate the most useful information in a form that could be understood and adopted by engineers in the context of engineering design and construction studies.

Since 1990, there have continued to be occasional publications indicating that land surface evaluation techniques are being used for engineering studies and that the available techniques have been developing rapidly (Hearn 1995*a*, *b*, 1997; Griffiths *et al.* 1996; Waller & Phipps 1996; Edwards 1997; Fookes 1997). Of particular importance are the changes in the remote sensing platforms that are able to provide data at scales that are suitable for engineering practice (e.g. Griffiths *et al.* 1994) with resolutions of ± 3 m now being commercially available. Aerial photographic interpretation continues to provide the basis of land surface evaluation in many engineering studies (Verstappen 1983; Lawrence *et al.* 1993). The modern techniques of analytical photogrammetry (Chandler & Moore 1989; Chandler & Brunsden 1995) and digital photogrammetry (Brunsden & Chandler 1996) using both aerial and terrestrial photography have added an important new dimension to the technique. The full potential for using digitized airborne multispectral imaging systems for application to civil engineering projects has yet to be realized despite the fact that it is universally recognized as a fundamental tool in military engineering and has been for many years.

In conjunction with these developments the new facilities available for data handling and analysis associated with the use of computers is comprehensive. The most immediately important development is the Geographical Information System (GIS). The capability of GIS to handle, analyse and reproduce large volumes of spatial data, and then to permit revision and updating of this data, is only just beginning to impinge on the world of civil engineering. It will provide a critical future development in promoting the scope and value of land surface evaluation in engineering practice.

The unequivocal endorsement of conceptual ground modelling, typically by the application of land surface evaluation techniques, presented appropriately in the First Glossop Lecture (Fookes 1997), pointed to the fundamental principles set out in the 1982 Working Party Report. Professor Fookes stated that field and laboratory site investigation work normally should be a carefully planned, largely confirmatory study rather than an exploratory exercise. The cost of such field and laboratory work and the type and range of techniques applied must be pertinent to the project development. Their deployment typically should be concentrated where they are needed to improve the existing intelligence database and to address 'problem' or 'unique' areas identified from the conceptual modelling. 'Type' areas typically only require confirmatory investigation work and the installation of

monitoring equipment to establish critical process parameters or seasonal variation in ground conditions that have been identified as being of engineering significance.

Since the First Working Party reported, therefore, land surface evaluation techniques have continued to develop, although in the UK this has mainly been in development planning. However, work carried out over the last two decades has demonstrated that land surface

Table 1. *Names and affiliation of contributors*

Professor R. Allison, Department of Geography, University of Durham

G. P. Birch, Consultant Engineering Geologist, Sevenoaks, Kent

Professor R. Blong, Natural Hazards, Research Centre, Macquarie University, Australia

Professor Brunsden, Emeritus Professor of Physical Geography, King's College, London

Dr P. Carey, School of Earth and Environmental Sciences, University of Greenwich

J. H. Charman, Consultant Engineering Geologist, Milford, Guildford, Surrey

Dr J. H. Chandler, Department of Civil and Building Engineering, Loughborough University

R. J. G. Edwards, Earth Science Partnership, Leatherhead, Surrey

Dr C. N. Edmonds, Peter Brett Associates, Reading, Berkshire

Professor P. G. Fookes, Consultant Engineering Geologist, Winchester, Hampshire

D. J. French, W. S. Atkins Consultants Ltd, Epsom, Surrey

Dr J. S. Griffiths, Department of Geological Sciences, University of Plymouth, Devon

N. E. Harrison, W. S. Atkins Consultants Ltd, Epsom, Surrey

Dr G. J. Hearn, Scott Wilson Kirkpatrick & Co Ltd, Basingstoke, Hampshire

I. Hodgson, Scott Wilson Kirkpatrick & Co Ltd, Basingstoke, Hampshire

Dr G. Humphreys, School of Earth Sciences, Macquarie University, Australia

E. M. Lee, Department of Marine Sciences and Coastal Management, University of Newcastle, Newcastle upon Tyne

Professor D. K. C. Jones, Department of Geography and Environment, London School of Economics, London

Dr R. P. Martin, Geotechnical Engineering Office, Civil Engineering Department, Government of the Hong Kong Special Administrative Region

Dr R. Moore, Sir William Halcrow and Partners, Birmingham

Dr C. P. Nathanail, School of Chemical Environmental and Mining Engineering, Nottingham University, Nottinghamshire

Dr D. N. Petley, Department of Geography, University of Durham

P. J. Phipps, Mott MacDonald, Croydon, Surrey

W. Rankin, Mott MacDonald, Croydon, Surrey

C. F. Sakalas, High-Point Rendel, Birmingham

D. T. Shilston, W. S. Atkins Consultants Ltd, Epsom, Surrey

M. Sweeney, BP-Amoco Exploration, London

S. Woddy, Scott Wilson Kirkpatrick & Co Ltd, Basingstoke, Hampshire

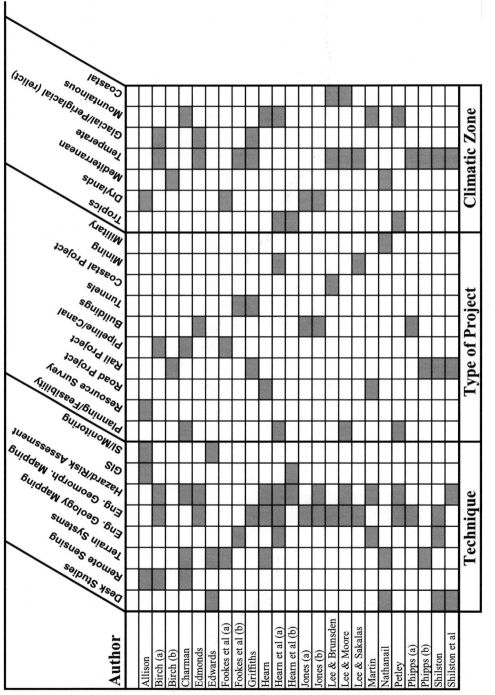

Fig. 1. Case studies index.

evaluation, particularly when used in conjunction with comprehensive and effective desk studies, is one the most cost-effective primary investigation techniques. One of its main strengths is in establishing conceptual ground models that can be used for engineering purposes, as described by Fookes (1997). In this role, desk studies developed around a land surface evaluation framework facilitate site investigation planning and enhance the value of subsurface investigation work. The techniques, however, have been shown to have an ongoing role during construction to assist in the interpretation of ground conditions. It is a matter of concern that the updated version of the *Code of Practice for Site Investigation* (British Standards Institution 1999) failed to recognize or endorse this perspective by again placing the emphasis on actual ground investigations using exploratory holes, geophysics, and field and laboratory testing, in the same manner as BS5930: 1981.

There are some important new developments in techniques for data collection, compilation and analysis. In land surface evaluation these will not replace traditional methods but will dramatically enhance the facilities for information storage, manipulation, revision and presentation. This is an opportune time, therefore, to undertake a review of the techniques of land surface evaluation in the context of engineering practice and to compile working examples of good practice in order to promote new standards of excellence for the next decade.

The Second Working Party Report

The brief for the Second Working Party, set up in January 1997, was to provide a revision of the First Working Party Report incorporating the changes that had occurred since 1982 and to provide a guide for best practice in the application of land surface evaluation techniques.

Following extensive discussions it was decided that the wide range of techniques that had developed since the First Working Party precluded the development of a 'manual of good practice'. The approach adopted for the report, therefore, was to invite state-of-the-art practitioners in land surface evaluation to contribute papers on techniques and engineering or planning case studies. Thus the report is not intended to be an instruction book that provides methodologies for applying all the various techniques because these are available in a range of specialist publications. The aim is to outline the available methods of land surface evaluation, identify where to find out details of how to use the techniques, and to provide examples of their application and output for a range of engineering and environmental situations.

The Second Working Party Report is divided into four sections:

Section 1: An introduction on the development of land surface evaluation for engineering practice.

Section 2: A series of papers on techniques to be used in land surface evaluation.
Section 3: A collection of land surface evaluation case studies from all types of engineering situations and environments.
Section 4: A conclusion on the future role of land surface evaluation in engineering practice.

The names and affiliation of all contributory authors are presented in Table 1.

To facilitate use of the Report, Figure 1 presents a cross-tabulation of techniques, engineering situations and climatic or environmental zones against the authors of papers presented in Section 3. References quoted by the authors can be found at the end of each paper.

References

AITCHISON, G. D. & GRANT, K. 1967. The P.U.C.E. programme of terrain description, evaluation and interpretation for engineering purposes. *In*: *Proceedings of the Fourth Regional Conference in Africa on Soil Mechanics and Foundation Engineering* (Cape Town), **1**, 1–8.

AITCHISON, G. D. & GRANT, K. 1968. Proposals for the application of the P.U.C.E. programme of terrain classification and evaluation to some engineering problems. *In*: *Symposium on Terrain Evaluation for Engineering, Fourth Conference of the Australian Road Research Board*, **4**, 1648–1660.

ANON. 1972. The preparation of maps and plans in terms of engineering geology. Report by the Geological Society Engineering Group Working Party. *Quarterly Journal of Engineering Geology*, **5**, 295–382.

ANON. 1982. Land surface evaluation for engineering practice. *Quarterly Journal of Engineernig Geology*, **15**, 265–316.

BECKET, P. H. T. & WEBSTER, R. 1969. *A review of studies on terrain evaluation by the Oxford-MEXE-Cambridge group*. MEXE Report **1123**, Christchurch, Hants.

BELCHER, D. J. 1948. The engineering significance of landforms. *Bulletin of the Highways Research Board*, **13**.

BRINK, A. B. A. & PARTRIDGE, T. C. 1967. Kyalami land system: an example of physiographic classification for the storage of terrain data. *In*: *Proceedings of the Fourth Regional Conference for Africa on Soil Mechanics and Foundation Engineering* (Cape Town), **1**, 9–14.

BRINK, A. B. A., PARTRIDGE, T. C., WEBSTER, R. & WILLIAMS, A. A. B. 1968. Land classification and data storage for the engineering use of natural materials. *In*: *Proceedings of the Fourth Conference of the Australian Road Research Board*, 1624–1647.

BRITISH STANDARDS INSTITUTION. 1957. *Site Investigations*. British Standard Code of Practice CP2001. The Council for Codes of Practice, British Standards Institution, London.

BRITISH STANDARDS INSTITUTION. 1981. *Code of Practice for Site Investigations*. BS5930: 1981. British Standards Institution, London.

BRITISH STANDARDS INSTITUTION. 1999. *Code of Practice for Site Investigations*. BS5930: 1999. British Standards Institution, London.

BROOK, D. & MARKER, B. R. 1987. Thematic geological mapping as an essential tool in land-use planning. *In*: CULSHAW, M. G., BELL, F. G., CRIPPS, J. C. & O'HARA, M. (eds) *Planning and Engineering Geology*. Geological Society, London, Engineering Geology Special Publications, **4**, 211–214.

BRUNSDEN, D. & CHANDLER, J. H. 1996. The development of an episodic landform change model based upon the Black Ven mudslide 1946–1995. *In*: ANDERSON, M. J. & BROOKES, S. M. (eds) *Advances in Hillslope Processes*. J. Wiley & Sons, Chichester, **2**, 869–896.

BRUNSDEN, D., DOORNKAMP, J. C., FOOKES, P. G., JONES, D. K. C. & KELLY, J. M. H. 1975. Large-scale geomorphological mapping techniques and highway engineering. *Quarterly Journal of Engineering Geology*, **8**, 227–254.

CHANDLER, J. H. & BRUNSDEN, D. 1995. Steady state behaviour of the Black Ven mudslide: the application of archival photogrammetry to studies of landform change. *Earth Surface Processes and Landforms*, **20**, 255–275.

CHANDLER, J. H. & MOORE, R. 1989. Analytical photogrammetry: a method for monitoring slope instability. *Quarterly Journal of Engineering Geology*, **22**, 97–110.

CHARMAN, J. H. & GRIFFITHS, J. S. 1993. Terrain evaluation methods for predicting relative hazard from mass movement, fluvial erosion and soil erosion in the developing world. *In*: MERRIMAN, P. A. & BROWITT, C. W. A. (eds) *Natural Disasters: Protecting Vulnerable Communities*. Thomas Telford, London, 167–183.

CLARK, A. R. & JOHNSON, D. K. 1975. Geotechnical mapping as an integral part of site investigation two case studies. *Quarterly Journal of Engineering Geology*, **8**, 211–224.

DEARMAN, W. R. 1987. Land evaluation and site assessment: mapping for planning purposes. *In*: CULSHAW, M. G., BELL, L. G., CRIPPS, J. C. & O'HARA, M. (eds) *Planning and Engineering Geology*. Geological Society, London, Engineering Geology Special Publications, **4**, 195–202.

DEARMAN, W. R. & FOOKES, P. G. 1974. Engineering geological mapping for civil engineering practice in the United Kingdom. *Quarterly Journal of Engineering Geology*, **7**, 223–256.

DEPARTMENT OF THE ENVIRONMENT. 1990. *Development on Unstable Ground*. Planning Policy Guidance PPG14, HMSO, London.

DEPARTMENT OF THE ENVIRONMENT. 1991. *Applied Earth Science Mapping*. HMSO, London, pp. 41.

DEPARTMENT OF THE ENVIRONMENT. 1995. *Development on Unstable Ground: Landslides and Planning*. Planning Policy Guidance PPG14 Annex 1 HMSO, London.

DOORNKAMP, J. C. (ed.) 1988. *Applied Earth Science Background: Torbay*. Report to the Department of the Environment.

DOORNKAMP, J. C., BRUNSDEN, D., JONES, D. K. C., COOKE, R. U. & BUSH, P. R. 1979. Rapid geomorphological assessments for engineers. *Quarterly Journal of Engineering Geology*, **12**, 189–204.

DOWLING, J. W. F. 1968. The classification of terrain for road engineering purposes. Conference on Civil Engineering Problems Overseas, Session V, Communications Paper 16, Institution of Civil Engineers, London.

DOWLING, J. W. F. & BEVAN, P. J. 1969. Terrain evaluation for road engineers in developing countries. *Journal of the Institution of Highway Engineers*, **16**.

EDWARDS R. J. G. 1968. *The use of aerial photographic interpretation for the analysis of engineering problems in the field of engineering geology*. MSc Thesis, Imperial College, University of London.

EDWARDS R. J. G. 1997. A review of the hydrogeological studies of the Cardiff Bay Barrage. *Quarterly Journal of Engineering Geology*, **30**, 49–62.

FOOKES, P. G. 1997. Geology for engineers: the geological model, prediction, and performance. *Quarterly Journal of Engineering Geology*, **30**, 293–424.

FOOKES, P. G., SWEENEY, M, MANBY, C. N. D. & MARTIN, R. P. 1985. Geological and geotechnical engineering aspects of low-cost roads in mountainous terrain. *Engineering Geology*, **21**, 1–152.

FORSTER, A., HOBBS, P. R. N., WYATT, R. J. & ENTWISLE, D. C. 1987. Environmental geology maps for Bath and the surrounding area for engineers and planners. *In*: CULSHAW, M. G., BELL, F. G., CRIPPS, J. C. & O'HARA, M. (eds) *Planning and Engineering Geology*. Geological Society, London, Engineering Geology Special Publications, **4**, 221–236.

GRIFFITHS, J. S. & HEARN, G. J. 1990. Engineering geomorphology: a UK perspective. *Bulletin of the International Association of Engineering Geology*, **42**, 39–44.

GRIFFITHS, J. S. & MARSH, A. H. 1986. BS5930: the role of geomorphological and geological techniques in a preliminary site investigation. *In*: HAWKINS, A. B. (ed.) *Site Investigation Practice*. Geological Society, London, Engineering Geology Special Publications, **2**, 261–271.

GRIFFITHS, J. S., FOOKES, P. G., HARDINGHAM, A. D. & BARSBY, R. D. 1994. Geotechnical soils mapping for construction purposes in Central Saudi Arabia. *In*: FOOKES, P. G. & PARRY, R. H. G. (eds) *Engineering Characteristics of Arid Soils*. Balkema, Rotterdam, 69–85.

GRIFFITHS, J. S., BRUNSDEN, D., LEE, E. M. & JONES, D. K. C. 1996. Geomorphological investigations for the Channel Tunnel portal and terminal. *The Geographical Journal*, **161**, 275–284.

GEOMORPHOLOGICAL SERIES LTD. 1987. *Review of research into landsliding in Great Britain*. Department of the Environment Volume Report, **14**.

HEARN, G. J. 1995a. Engineering geomorphological mapping and open-cast mining in unstable mountains – a case study. *Transactions of the Institution of Mining and Metallurgy*, Section A, **104**, A1–A17.

HEARN, G. J. 1995b. Landslide and erosion hazard mapping at Ok Tedi copper mine, Papua New Guinea. *Quarterly Journal of Engineering Geology*, **28**, 47–60.

HEARN, G. J. 1997. *Principles of low cost road engineering in mountainous regions, with special reference to the Nepal Himalaya*. Overseas Road Note 16, Transportation Research Laboratory, Crowthorne.

JONES, D. K. C., BRUNSDEN, D. & GOUDIE, A. S. 1983. A preliminary geomorphological assessment of part of the Karakoram highway. *Quarterly Journal of Engineering Geology*, **16**, 331–356.

LAWRENCE, C. J. 1978. *Terrain Evaluation in West Malaysia – Part 2*. Transportation Road Research Laboratory Report SR 378, TRRL, Crowthorne.

LAWRENCE, C. J., BYARD, R. J. & BEAVEN, P. J. 1993. *Terrain Evaluation Manual*. Transportation Research Laboratory, State-of-the-Art Review 7, HMSO, London.

LEE, E. M., DOORNKAMP, J .C., BRUNSDEN, D. & NOTON, N. H. 1991. *Ground movement in Ventnor, Isle of Wight*. Report to the Department of the Environment.

MEXE. 1969. *Field trials of a terrain classification system.* MEXE Report **995**, Christchurch, Hants.

MITCHELL, C. W. 1973. *Terrain Evaluation.* Longmans, London.

NATIONAL INSTITUTE FOR ROAD RESEARCH. 1971. *The production of soil engineering maps for roads and the storage of materials data.* NIRR Technical Recommendations for Highways, **2**.

PERRIN, R. M. S. & MITCHELL, C. W. 1969. *An appraisal of physiographic units for predicting site conditions in arid areas.* MEXE Report **1111**, Vols I & II, Christchurch, Hants.

SMITH, A. & ELLISON, R. A. 1999. Applied geological maps for planning and development: a review of examples for England and Wales, 1983 to 1996. *Quarterly Journal of Engineering Geology*, **32**, Supplement, May 1999.

VERSTAPPEN, H. TH. 1983. *Applied Geomorphology: Geomorphological Surveys for Environmental Development.* Elsevier, Amsterdam.

VERSTAPPEN, H. TH. & VAN ZUIDAN, R. A. 1968. *Photointerpretation.* ITC, Delft.

WALLACE EVANS LTD. 1994. *The presentation of earth science information for planning, development and conservation – illustrated by a study of the Severn Levels.* Summary Report to the Department of the Environment.

WALLER, A. W. & PHIPPS, P. 1996. Terrain systems mapping and geomorphological studies for the Channel Tunnel rail link. *In*: CRAIG, C. (ed.) *Advances in Site Investigation Practice.* Thomas Telford, London, 25–38.

Section 2

Techniques in Land Surface Evaluation

Terrain measurement using automated digital photogrammetry

J. H. Chandler

Department of Civil and Building Engineering, Loughborough University, Leicestershire, UK

Introduction

Photogrammetry has traditionally provided a means of generating three-dimensional spatial data to represent terrain surfaces, which complements traditional ground-based surveying methods. Although techniques such as airborne laser scanning (Lohr 1998) and synthetic aperture radar (Hogg *et al.* 1993; Vencatasawamy *et al.* 1998) have developed, photogrammetry remains the primary method of generating topographic maps (Wolf 1983; Capes 1998). One important advantage of photogrammetry is the flexibility of scale that allows application to imagery acquired from ground, air and space. Indeed, a new generation of high (i.e. 1 m) resolution satellite sensors (Capes 1998) is likely to further increase the potential applications of photogrammetry. Despite many advantages, there have been several problems with the application of photogrammetry using traditional methods. Most significantly, there was the requirement to use an expensive and complex photogrammetric stereo-plotter. This ensured that the measurement process was slow and generally required the skills of an experienced operator, particularly if results of the highest accuracy were to be obtained.

Rapid developments in computing hardware and software have allowed the science of photogrammetry to develop rapidly during the last ten years (Gruen 1994; Atkinson 1996; Greve 1996). These developments have radically eased many of the problems and limitations associated with traditional analogue instrumentation. Use of a purely numerical or analytical solution provides flexibility, which assists in two important ways. Satellite imagery, oblique aerial photography and ground-based imagery can be used, in addition to the more traditional vertical aerial perspective. Similarly, imagery acquired using a variety of non-photogrammetric cameras can be considered to be of value for spatial measurement. The most recent advance, known as digital photogrammetry, now allows part of the measurement process to be fully automated. This significant development ensures that photogrammetry represents an even more versatile and efficient method of deriving dense digital elevation models (DEMs) to represent the geometric characteristics of land and terrain surfaces.

Digital image acquisition

Digital photogrammetric methods rely upon the use of a digital image instead of the more traditional analogue contact diapositive. The digital image comprises a large array of pixels, each representing a particular colour or grey-scale value to form the overall image. Space-borne imaging systems use push-broom sensors to create the digital representation. For large and medium scales, digital imagery can be obtained directly using a digital camera (Koh & Edwards 1996). Although digital camera technology provides instant and appropriate imagery, direct digital/vertical/aerial image acquisition is rarely used by the photogrammetric community (Maas & Kersten 1997). This is partly because digital cameras remain expensive, but, more crucially, such cameras only generate images at comparatively low resolution (e.g. the Kodak DCS460 camera costs £15 000 and provides an image of only 3000×2000 pixels). The emulsions used in conventional aerial films represent an extremely efficient means of storing image data at resolutions equivalent to $50\,000 \times 50\,000$ pixels. Currently the most cost-effective means of obtaining digital imagery involves a hybrid approach (Helava 1988) in which the conventional analogue photograph is converted into digital form using some form of scanning process. Scanning options and costs vary widely, but to enable simplified processing using full-format aerial images, a purpose-built geometrically stable scanner should be used, (Warner *et al.* 1996). These are expensive to buy, but bureau scanning services provide a cheaper alternative if production volume is low.

It should be remembered that converting a photograph into digital form requires significant volumes of file storage, which can cause serious system management issues. For example, a normal black and white aerial photograph (dimensions 230×230 mm) scanned at 25 micron resolution (1016 dpi) with 256 grey levels will generate an image consisting of 9200×9200 pixels,

From: GRIFFITHS, J. S. (ed.) *Land Surface Evaluation for Engineering Practice*. Geological Society, London, Engineering Geology Special Publications, **18**, 13–18. 0267-9914/01/$15.00 © The Geological Society of London 2001.

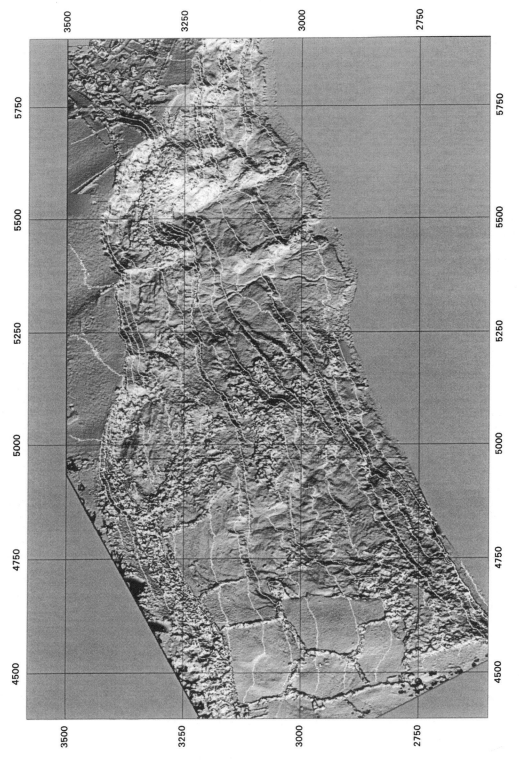

Fig. 1. Slope shaded 1 m resolution DEM with 10 m contours overlain: Black Ven mudslide complex.

occupying 81 Mb of disk space. A minimum of two images is needed to carry out three-dimensional measurement and if colour images are scanned then these files need to increase in size by a factor of three. The management issues raised by working with digital imagery are described more fully by Colomer & Colomina (1994).

Image measurement and automatic DEM generation

One of the important advantages of digital photogrammetry is the replacement of the complex, specialized and expensive photogrammetric plotter with a more general, and hence far cheaper, digital computer. Machines running the UNIX operating system have been widely used in the recent past, but PC-based platforms are becoming ever more powerful and capable of manipulating the large images required. A competent UNIX machine can be purchased for costs as little as £3000, a PC for less. Appropriate photogrammetric software is required but competition for this market and the influence of software packages designed for remote sensing applications have reduced costs significantly (e.g. Erdas Imagine/OrthoMax and OrthoBase, PCI/EASI-PACE, R-WEL/Desktop Mapping System, VirtuoZo). Prices vary widely, currently within a range between £4000 and £40 000.

The second and perhaps most significant advantage of digital photogrammetry is the potential to automate various aspects of the measurement process. The essential capability relevant for land surface measurement is the automated measurement of digital elevation models from an overlapping stereo-pair, which is now both practicable and in an advanced state of development. Such automation is based upon sophisticated image correlation or image matching techniques that automatically identify and measure common image patches appearing on two overlapping digital images. Once matched, these two image measurements are transformed into object coordinates using established photogrammetric methods, and the process repeated. With appropriate hardware and software, this cycle can recur at speeds in excess of 100 points per second and so very dense and consequently accurate DEMs can be generated. In most packages, the derived DEM is in the form of a regular grid of elevation estimates draped over the desired area. This ability to measure very dense and regular grid DEMs has instigated a return to grid-based methods of manipulating and presenting height information. This will perhaps reverse the trend towards using DEM processing methods based upon the Delauney triangulation (Petrie & Kennie 1990). This algorithm efficiently creates a surface from a limited sample of terrain elevations in which the break-line assumes great significance.

Application

Photogrammetric software packages, available commercially, have been developed for application using either stereo-satellite imagery (i.e. SPOT) or traditional vertical aerial photography. One comparatively routine application of automated digital photogrammetry is illustrated by the work of Brunsden & Chandler (1996) who acquired a new epoch of vertical aerial photography (photo scale 1:4000) of the Black Ven landslide, Dorset, UK, in March 1995. Automated methods were used to generate a DEM consisting of 1 000 000 points within an area of 1250 × 800 m, a sampling density of one point every metre (Fig. 1). Once a DEM had been created, it was possible to create contours (Fig. 1), cross-sections and orthophotos (Fig. 2). The orthophoto is particularly valuable for Earth scientists because it combines the interpretative capabilities of the original aerial photograph with the positional relevance of a map. The 1995 DEM of the Black Ven system was used to update a sequence of five lower resolution manually measured DEMs, which represented the morphology of the mudslides every ten years since 1946 (Chandler & Cooper 1989). The improved spatial resolution in combination with climatic and landslide incidence data allowed the revision of an evolutionary model (Brunsden & Chandler 1996).

Although application using vertical aerial photographs will always remain most important, it is significant that automated digital photogrammetry can be applied to both ground-based and terrestrial imagery. To ensure that the automated DEM extraction software remains successful, it is necessary to introduce an extra stage in the photogrammetric processing and although this is not documented in software user manuals, it is possible (Chandler 1999). One application of this approach is reported by Pyle et al. (1997) and involved the creation of DEMs representing riverbanks using ground-based oblique photography. The objective was to compare successive DEMs in order to map the spatio-temporal pattern of bank erosion. The precision of automatically generated DEMs was ±12 mm and, significantly, the locations where individual clasts had been removed from the gravel riverbank could be identified. The same methods have been used to quantify the three-dimensional form of exposed and subaerial riverbed gravels, both in natural riverbed gravels (Butler et al. 1998) and in a flume (Stojic et al. 1998).

Limitations

When considering using digital photogrammetric techniques, it is important to consider some of the requirements and limitations associated with the method. Although not theoretically essential, introducing photocontrol points into the object space eases photogrammetric processing significantly and should always be

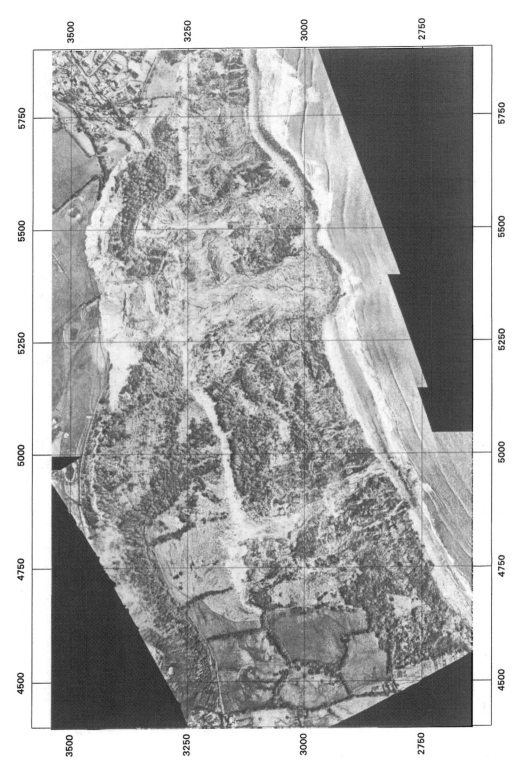

Fig. 2. Orthophoto (true map) of Black Ven mudslide complex.

considered. Photo-control points are simply recognizable features that appear on the photography, and are situated at known geographical locations. Such points provide the means to establish the transformation from measurements on the image, to positions in the desired ground coordinate system. The technology available to carry out photo-control surveys has advanced in recent years (Lane *et al*. 1998). For surveys of small areas using terrestrial photogrammetry, the modern total station is most effective, particularly if combined with an intersection method of surveying. Over larger areas and using vertical aerial imagery, the Global Positioning System (GPS) (Schofield 1994) is particularly appropriate, mainly because GPS does not rely upon maintaining 'line of sight' between points.

Although automation afforded by digital photogrammetry provides a distinct advantage, such automation can also be interpreted as a weakness. The software will always produce some form of surface, but this may not necessarily represent the surface that the user requires. If, for example, the parameters used to control the derivation of the DEM are inappropriate, then erroneous surfaces will be generated (Smith *et al*. 1996). In vegetated areas, the surface generated automatically will represent the tops of all visible vegetation and not the underlying ground surface, which the user may perhaps require. Automated DEM acquisition is possibly best suited to situations in which there is either a lack of vegetation or where vegetation is homogenous in terms of foliage height above the true ground surface. Sharp discontinuities in the land surface can also downgrade final surface representation, particularly where resolution of the sampling points is low. It may then become necessary to measure the three-dimensional characteristics of such discontinuities directly, using break-lines (Petrie & Kennie 1990).

Many packages provide the opportunity to assess the accuracy of points measured automatically using stereo-superimposition (Smith *et al*. 1996). This practice is valuable but although such packages provide the opportunity to edit and correct erroneous height estimates, this process can become impracticable if the density of generated data is high.

Conclusion

Digital elevation models can now be generated automatically using digital photogrammetry. Such automation has allowed the density of data used to represent terrain morphology to be increased by a factor of over 100 compared to manual photogrammetric measurement methods. This development is of distinct value to terrain measurement because it replaces the tedious and expensive traditional manual procedures and generates a far higher resolution DEM.

Developments in software implementing digital photogrammetric methods are significant and readers are encouraged to consider using such packages. The software is also 'user-friendly', which allows the inexperienced novice user to obtain results. Despite this, it is perhaps important to conclude with several cautionary comments. First, it is necessary to become familiar with the procedures briefly outlined in this paper. More guidance is provided by photogrammetry textbooks (Wolf 1983; Atkinson 1996; Greve 1996) and other papers (i.e. Chandler 1999). It is easy to become overly ambitious in terms of the size and number of DEMs that can be generated in a project. It is prudent to start with a small area and extend only when experience has been gained. Finally, it must be recognized that it takes considerable care and some expertise to generate truly accurate surfaces on a routine basis. It is always essential to assess the accuracy of generated surfaces using independent methods.

References

ATKINSON, K. B. 1996. *Close Range Photogrammetry and Machine Vision*. Whittles, Caithness.

BRUNSDEN, D. & CHANDLER, J. H. 1996. The development of an episodic landform change model based on the Black Ven mudslide 1946–95. *In*: ANDERSON, M. J. & BROOKS, S. M. (eds) *Advances in Hillslope Processes*. John Wiley & Sons, Chichester, **2**, 869–898.

BUTLER, J., LANE, S. N. & CHANDLER, J. H. 1998. DEM Quality assessment for surface roughness characterisation using close-range photogrammetry. *Photogrammetric Record*, **16**(92), 271–291.

CAPES, R. 1998. Developments in earth observation – how new space-borne sensors will affect mapping. *Surveying World*, **6**(6), 24–27.

CHANDLER, J. H. 1999. Effective application of automated digital photogrammetry for geomorphological research. *Earth Surface Processes and Landforms*, **24**, 51–63.

CHANDLER, J. H. & COOPER, M. A. R. 1989. The extraction of positional data from historical photographs and their application in geomorphology, *Photogrammetric Record*, **13**(73), 69–78.

COLOMER, J. L. & COLOMINA, I. 1994. Digital photogrammetry at the Institut Cartogràfic de Catalunya. *Photogrammetric Record*, **14**(84), 943–956.

GREVE, C. 1996. *Digital Photogrammetry: an addendum to the manual of photogrammetry*. American Society of Photogrammetry and Remote Sensing, Bethesda.

GRUEN, A. 1994. Digital close-range photogrammetry- progress through automation. *International Archives of Photogrammetry and Remote Sensing*, **5**, 122–135.

HELAVA, U. V. 1988. On system concepts for digital automation. International Archives of Photogrammetry and Remote Sensing, 27/2, Kyoto, 171–190.

HOGG, J., McCORMACK, J. E., ROBERTS, S. A. & GAHEGAN, CHANDLER, J. H.M. N. 1993. Automated derivation of stream channel networks and selected catchment characteristics from digital elevation models. *In*: P. M. MATHER (ed.) *Geographical Information Handling – Research and Applications*. John Wiley & Sons, Chichester, 207–235.

KOH, A. & EDWARDS, E. 1996. *Integrating GPS data with Fly-on-demand digital imagery for coastal zone management.* Association of Geographic Information 1996 Conference Proceedings, 6.1.1–6.1. 5.

LANE, S. N., CHANDLER, J. H. & RICHARDS, K. S. 1998. Landform monitoring, modelling and analysis: landform in geomorphological research. *In*: LANE, S. N., RICHARDS, K. S. & CHANDLER, J. H. (eds) *Landform Monitoring, Modelling and Analysis*. John Wiley & Sons, Chichester, 1–17.

LOHR, U. 1998. Digital elevation models by laser scanning, *Photogrammetric Record*, **16**(91), 105–109.

MAAS, H.-G. & KERSTEN, T. 1997. Aero-triangulation and DEM/Orthophoto generation from high-resolution still-video imagery. On the potential of digital cameras onboard an aircraft. *Photogrammetric Engineering and Remote Sensing*, **63**(9), 1079–1084.

PETRIE, G. & KENNIE, T. J. M. (eds). 1990. *Terrain Modelling in Surveying and Civil Engineering*. Whittles, Caithness.

PYLE, C. J., RICHARDS, K. S. & CHANDLER, J. H. 1997. Digital photogrammetric monitoring of river bank erosion. *Photogrammetric Record*, **15**(89), 753–763.

SCHOFIELD, W. 1994. *Engineering Surveying*. Butterworth-Heinemann, Oxford.

SMITH, M. J., SMITH, D. G. & WALDRAM, D. A. 1996. Experiences with analytical and digital stereoplotters. *Photogrammetric Record*, **15**(88), 519–526.

STOJIC, M., CHANDLER, J. H., ASHMORE, P. & LUCE, J. 1998. The assessment of sediment transport rates by automated digital photogrammetry. *Photogrammetric Engineering and Remote Sensing*, **645**, 387–395.

VENCATASAWAMY, C. P., CLARK, C. D. & MARTIN, R. J. 1998. Landform and lineament mapping using radar remote sensing. *In*: LANE, S. N., RICHARDS, K. S. & CHANDLER, J. H. (eds) *Landform Monitoring, Modelling and Analysis*. John Wiley & Sons, Chichester, 165–194.

WARNER, W. S., GRAHAM, R. W. & READ, R. E. 1996. *Small Format Aerial Photography*. Whittles, Caithness.

WOLF, P. R. 1983. *Elements of Photogrammetry*. McGraw Hill, Singapore.

Desk studies

J. H. Charman

Consultant Engineering Geologist, Milford, Guildford, Surrey, UK

Introduction

In many, if not most, instances of civil engineering or building projects that have suffered time or cost over-runs or have required premature remedial works, the reason can be attributed to geotechnical problems as a result of inadequate planning or poor interpretation of site investigation (Site Investigation Steering Group 1993). The difficulties experienced in civil engineering projects which can be related to the ground conditions occur either as a result of unawareness or as a result of failure to grasp the implications of a certain set of ground conditions on the proposed engineering design (Fookes 1997a). The site ground conditions pertaining at the present are the result of a long history of geological pro-cesses, from global tectonics, through climatic change to relatively recent landscape-forming processes. Major impact may also have been caused by man's historical usage of the site. Much of this history of usage can be derived by the specialist from existing scientific records and published maps. The desk study is a fundamental first step in any site investigation programme. Its pur-pose is to access published information and other avail-able records pertinent to the region, area and immediate environs of the project development site. This would in-clude an investigation of geology, geomorphology, aerial photographs and other archival data.

The recently published updated British Standard *Code of Practice for Site Investigation*, BS5930:1999, is unequivocal in stating that Stage 1 of a site investigation comprises a desk study and site reconnaissance and it should be undertaken at the start of every investigation (British Standards Institution 1999).

Objectives

The main objective of the desk study is to carry out a preliminary assessment of the ground conditions based on existing information, to use this as a basis for a reconnaissance of the site and to plan the scope of the ground investigation. This objective can be achieved at relatively low cost and at an early stage in the investigation programme. If carried out by a suitably qualified and competent engineering geologist or geo-technical engineer, it is arguably the most cost-effective stage of the investigation.

Sources of information

Information that aids the implementation of the ground investigation and the overall project includes planning and statutory restrictions, land ownership and access considerations, and access to utilities. These sources are given in BS5930:1999 and are not considered further here. This paper deals with those sources relevant to the evaluation of the terrain and, thus, determination of the ground conditions and sources of construction mate-rials. Sources of information include published maps and memoirs, records from national and local govern-ment archives, aerial photographs, scientific papers and records from other development activity in the area. Of increasing significance is the availability of informa-tion on the World Wide Web and references to useful web sites are included below.

For terrain evaluation *per se*, primary information includes topographic maps, aerial photographs, geologi-cal maps and soils maps. These all provide base data that allow interpretation by the specialist. Secondary infor-mation such as reports and archival material may already include some form of interpretation or assessment.

In developing countries the available information is largely primary. In the UK and other developed coun-tries, many sources of secondary information also exist. The sources of information in the UK are comprehen-sively described in Perry & West (1996). A summary of the sources of primary information is provided below.

Topographic information

Current mapping information in the UK is available from the Ordnance Survey (www.ordsvy.gov.uk) at vari-ous scales in both hard copy and digital form. Use-ful maps for terrain evaluation are the 1:50 000 scale Landranger and 1:25 000 scale Pathfinder, Explorer and

From: GRIFFITHS, J. S. (ed.) *Land Surface Evaluation for Engineering Practice*. Geological Society, London, Engineering Geology Special Publications, **18**, 19–21. 0267-9914/01/$15.00 © The Geological Society of London 2001.

Outdoor Leisure series. For larger scales, urban mapping at 1:1250, rural mapping at 1:2500 and mountain and moorland mapping at 1:10 000 scale are available.

An examination of historic topographic maps is essential to trace, for example, previous industrial usage (Charman & Cooper 1987). It is worth remembering that such maps are a factual record of the ground surface at a single point in time. If, for example, a mine shaft is infilled it ceases to feature on subsequent map editions. Therefore, a sequential review of all historic maps for the site should be made. The most comprehensive collection is available at the British Library (www.bl.uk) and includes all Ordnance Survey map editions back to the 1800s. The British Library also contains many other antiquarian maps as does the library of the Royal Geographical Society (www.rgs.org).

In developing countries topographic maps are generally available from government survey departments. Security considerations often govern their availability and it is essential that a letter of authority is provided. Alternatively, for former colonies, maps are often available in the former colonial power.

Aerial photographs and satellite imagery

Aerial photographs and satellite imagery are particularly useful for the preparation of base maps and for terrain interpretation in undeveloped areas. They are often available for areas where other material is subject to security limitations and they can be used for those areas where access to the site is a problem. As with old map editions, where historic aerial photography is available the previous history of the site can be determined. In addition, some sources of satellite imagery provide an archive of regular images for a particular site going back over many years.

Stereoscopic study techniques used by the specialist remain the most effective tool for the initial assessment of the terrain. They allow identification of many features including landforms, drainage patterns, geological boundaries, unstable ground and solution features. Advances continue to be made in the use of colour, multispectral scanners and sensor systems.

The National Library of Air Photographs is probably the biggest UK collection and is held by the Royal Commission on the Historic Monuments of England (www.rchme.gov.uk), now part of English Heritage.

For satellite imagery the National Remote Sensing Centre (www.nrsc.co.uk) and the World Wide Web virtual library of remote sensing (www.vtt.fi/aut/rs/virtual/) are good places to begin a search for available images.

Geological information

Geological maps are available from the British Geological Survey (BGS) (www.bgs.ac.uk). A full catalogue and guide is available at the web site. Geological maps at a scale of 1:50 000 cover most of the UK and are accompanied by sheet memoirs. More detailed mapping at 1:10 000 scale is also available and often contains detailed field notes made by the mapping geologist.

The Geological Society has developed a new web site (www.geolsoc.org.uk) which allows the library catalogue to be searched and also acts as a gateway to other Earth science sites for world wide information. The library is one of the largest UK sources of geological information.

Borehole information

The BGS National Geological Records Centre has a comprehensive collection of site investigation and other borehole records, referenced to the relevant Ordnance Survey 1:10 000 map series. A charge is made for this information.

Soil survey maps

Pedological soil maps classify the soil in the top 1.5 m of the soil profile. The properties of this layer reflect the properties of the deeper soil and rock profile. This is particularly true in tropical areas of the world where residual soils dominate.

In many parts of the world the coverage of pedological soil maps is more extensive than geological map coverage. Pedological soil nomenclature is complex and several classification systems exist. The most widely used are those of the Food and Agricultural Organisation of the United Nations FAO/UNESCO (1988), the US Soil Conservation Service and Duchaufour (1982; for areas developed under French influence). Fookes (1997b) provides a summary of these systems as an aid to geotechnical classification in tropical regions.

Method of approach

The key to an efficient desk study is to focus clearly on the type of information that is potentially relevant to the project. For purposes of land surface evaluation it is essential that a base map is prepared. Therefore, the first step is to access the primary sources of information, i.e. topographic maps and aerial photographs. If these are unavailable then satellite imagery will provide a true-to-scale base map at a price. A good base map, at a scale to suit the project site and the project needs, provides the key to presenting and referencing other data sources. Once this map is prepared other sources of information can be meaningfully accessed and referenced. Geographical Information Systems (GIS) exist at various levels of sophistication. The simplest is a base map with annotated reference points and a series of derivative maps showing particular themes. Such themes may be factual, representing for example change in slope angle or depth of shallow mining across a site. The theme may be an interpretation, representing changes in foundation conditions across the site for shallow foundations. Data relevant to the reference points are best summarized in tabular form. The full range of thematic mapping exercises carried out in the United Kingdom based primarily on available information is discussed in Smith & Ellison (1999).

References

BRITISH STANDARDS INSTITUTION. 1999. BS5930: 1999 *Code of Practice for Site Investigation*. British Standards Institution, London.

CHARMAN, J. H. & COOPER, C. G. 1987. The Frindsbury area, Rochester: a review of historical data and their implication on subsidence in an urban area. *In*: CULSHAW, M. G., BELL, F. G., CRIPPS, J. C. & O'HARA, M. (eds) *Planning and Engineering Geology*. Geological Society, London, Engineering Geology Special Publications, **4**, 115–124.

DUCHAUFOUR, P. 1982. *Pedology, Pedogenesis and Classification* (English edition translated by T. R. Paton). George Allen and Unwin, London.

FOOD AND AGRICULTURAL ORGANISATION OF THE UNITED NATIONS. 1988. *Soil Map of the World Revised Legend*. World Soil Resources Report No. 60, FAO, Rome.

FOOKES, P. G. 1997a. Geology for engineers: the geological model, prediction and performance. *Quarterly Journal of Engineering Geology*, **30**, 294–424

FOOKES, P. G. (ed.) 1997b. *Tropical Residual Soils*. Geological Society Professional Handbooks, The Geological Society, London.

PERRY, J. & WEST. G. 1996. *Sources of information for site investigations in Britain*. TRL Report No. **192**, Transport Research Laboratory, UK.

SITE INVESTIGATION STEERING GROUP. 1993. *Site investigation in construction – Part 1: Without site investigation ground is a hazard*. Thomas Telford, London.

SMITH, A. & ELLISON, R. A. 1999. Applied geological maps for planning and development: a review of examples from England and Wales 1983 to 1996. *Quarterly Journal of Engineering Geology*, Supplement, **32**, S1–S44.

Engineering classification for environmental performance

J. H. Charman[1], P. J. Carey[2] & P. G. Fookes[3]

[1] Consultant Engineering Geologist, Milford, Guildford, Surrey, UK
[2] School of Earth and Environmental Sciences, University of Greenwich, Kent, UK
[3] Consultant Engineering Geologist, Winchester, Hampshire, UK

Purpose of investigations

The condition of natural soils and rocks reflects the impact of a historical sequence of geological processes (Fookes 1997), including plate tectonics, depositional environment, structural and diagenetic change. Climate influences the effect of the atmosphere in producing surface-related weathering and, in particular, the climatic changes of the Quaternary have significantly modified the properties of near-surface soils and rocks. Such modifications continue under present-day environmental conditions.

Man-made or engineering structures are much younger compared to the geological time-scale but they also undergo change as a result of the effect of the natural environment in which they have been placed. The impact of these changes is governed by the materials that have been used in the construction, the way in which they have been incorporated in the design, and the quality of the workmanship. The materials are not always suitable for the environment in which they have been placed. An example is the use of pre-cast concrete with shallow reinforcement cover in a coastal environment where it is subject to seawater attack.

It follows, therefore, that engineering structures may be placed in a wide variety of environments in which chemical and physical attack will vary from harsh to benign. For each environment the factors which may influence the rate of attack can be developed into a rating system which will allow the assessment of the condition of the structure and its constituent materials. This provides a framework for:

- development of a monitoring and maintenance programme;
- possible changes to more appropriate design;
- selection of more appropriate construction materials;
- possible improvements in workmanship.

The use of such rating systems in engineering geology is well established; for example, the assessment of 'stand up time' for underground excavations, the Rock Mass Classification (Bieniawski 1974) – has been in use for many years.

This paper gives an example of a classification of the environmental performance of natural and man-made structures within a coastal environment. A classification system is described which introduces a score rating for each of a series of identified factors contributing to the performance of the structure. The scheme should be readily adapted to other environments, if the variation in the geomorphological conditions and factors that influence them can be defined in a similar way.

Basis of the performance model

The approach starts by identifying the factors that are important to the performance of each coastal structure. These included geology, local coastal environment, age of structure, material of construction, condition and visual assessment of the frequency of repair. This formed the basis of a pro forma on which factors were either descriptive and given tick boxes or quantified through a rating index. Each of the rating categories is summarized below, but for points of detail the reader should refer to Figure 1.

Definition of structure

The subdivision of the structure into separate elements is critical and must be developed to suit the particular output required. For example, the assessment of a coastal section (Fig. 2a) may require subdivision into slope or beach types and individual coastal protection structures, e.g. groynes, sea-wall, etc. Assessment of an individual structure (Fig. 2b) will require subdivision into its components, e.g. piers, footway, balustrade, etc.

Geology

Changes in geological material imply changes in performance such as the rate of weathering, resistance to erosion and slope processes, and therefore must be rated when assessing a coastal section where the geology changes. The rating has been based on the fundamental

From: GRIFFITHS, J. S. (ed.) *Land Surface Evaluation for Engineering Practice*. Geological Society, London, Engineering Geology Special Publications, **18**, 23–28. 0267-9914/01/$15.00 © The Geological Society of London 2001.

GENERAL LOCATION AND REFERENCE DETAILS.

1. *STRUCTURE*		Tick box
1A	NATURAL COAST	
1A1	Building coast	
1A2	Eroding coast	
1B	MAN MADE COAST	
1B1	Groynes	
1B2	Wave return revetment	
1B3	Armoured breakwater	
1B4	Sea wall	
1B5	Promenade/path	
1B6	Other	

2. *GEOLOGY*	(Add description)		Score
2A1		Widely spaced (>600mm)	1
2A2	Strong (breaks with hammer)	Mod. spaced (200-600mm)	2
2A3		Closely spaced (<200mm)	4
2A4	Rock	Widely spaced (>600mm)	2
2A5	Weak (breaks in hand)	Mod. spaced (200-600mm)	4
2A6		Closely spaced (<200mm)	7
2A7	Soil	Fine	8
2A8		Coarse	10

3. *MATERIAL OF CONSTRUCTION*		Tick box
3A	STONE/MASONRY	
3B	CONCRETE	
3C	BITUMINOUS BOUND	
3D	TIMBER	
3E	METAL	
3F	BRICK	
3G	FILL	
3H	OTHER	

4. *AGE*		Score
4A	PRE-HISTORIC	1
4B	PRE 1900	4
4C	1900 -1940	6
4D	1940 -1960	7
4E	1960 -1980	8
4F	1980 -1990	9
4G	1990 - PRESENT	10

5. *ENVIRONMENT*		Score
5A	SUB-TIDAL	1
5B	INTERTIDAL	3-10
5C	SUPRA-TIDAL	9-3
5D	BACK BEACH	2
5E	INLAND	1

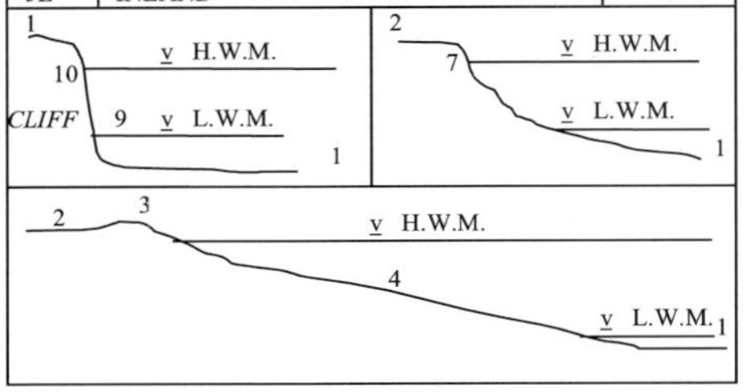

6. *COASTAL PROCESS*		Score
6A	LOW-LYING	1
6B	DEGRADED SLOPE	2-5
6C	ACTIVE DEGRADING SLOPE	5-9
6D	CLIFF	10

7. *FREQUENCY OF REPAIR*		Score
6A	NONE	1
6B	OCCASIONAL	2
6C	EVERY 10 YRS	3
6D	EVERY 5 YRS	8
6E	EVERY YEAR	10

8. *RATE OF DETERIORATION*		Score
7A	CHANGE OVER >100 YEARS	1
7B	CHANGE OVER 50 YEARS	2
7C	CHANGE OVER 10 YEARS	3
7D	CHANGE WITHIN 5 YEARS	8
7E	CHANGE WITHIN A YEAR	10

9. *CONDITION*		Score
8A	VERY GOOD	1
8B	GOOD	3
8C	ADEQUATE	5
8D	POOR	7
8E	VERY POOR	9
8F	FAILED	10

10. *PUBLIC SAFETY*		Score
9A	NO RISK	1
9B	LOW RISK	3
9C	MODERATE RISK	5
9D	HIGH RISK	8
9E	VERY HIGH RISK	10

Fig. 1. Ratings for classification pro formas.

(a)

GENERAL LOCATION AND REFERENCE DETAILS

1. STRUCTURE/CHAINAGE		0-150m	150-650m	650-655m	655-1200m	1200-1425m
1A	NATURAL COAST					
1A1	Building coast					
1A2	Eroding coast					
1B	MAN MADE COAST					
1B1	Groynes					
1B2	Wave return revetment					
1B3	Armoured breakwater					
1B4	Sea wall					
1B5	Promenade/path					
1B6	Other					

2. GEOLOGY					

3. MATERIAL OF CONSTRUCTION						
3A	STONE/MASONRY					
3B	CONCRETE					
3C	BITUMOUS BOUND					
3D	TIMBER					
3E	METAL					
3F	BRICK					
3G	FILL					
3H	OTHER					

4. AGE					

5. ENVIRONMENT					

6. COASTAL PROCESS					

7. FREQUENCY OF REPAIR					

8. RATE OF DETERIORATION					

9. CONDITION					

10. PUBLIC SAFETY					

11. RECOMMENDED ACTION					
REPAIR URGENTLY					
REPAIR RECOMMENDED					
MONITOR CONTINUOUSLY					
INSPECT ANNUALLY					
INSPECT AFTER 5 YRS					

(b)

Replace Section 1 (above) by the following:

1. STRUCTURE		SEAWARD PIER	LANDWARD PIER	RAIL WALL	STEPS	PATHWAY
1A	FOOTBRIDGE					

Fig. 2. Examples of classification pro formas: (**a**) for coastal section; (**b**) for individual structure.

properties of strength and discontinuity spacing in rock and particle size (coarse or fine) in soil (Fig. 1).

Local environment

The coastal environment has been subdivided into five divisions, which reflect the vulnerability to wave attack and exposure to salt weathering. The nature of the coastal profile influences the distribution of the subdivisions (Fig. 1), and the pro forma may be enhanced in some instances by the addition of a contoured plan.

Material of construction

As with the *in-situ* soil or rock, the materials of construction are fundamental to the durability of the structure. Comparison of the different elements of the pro forma will demonstrate those materials that perform best in any particular environment, and may lead to recommendations for the most suitable repair materials.

Age, rate of deterioration, frequency of repair and condition

These factors are used in combination to provide a measure of the durability of the structure or its fitness for purpose. The age of the structure is important because it measures the time over which the structure has been exposed. The rate of deterioration, for natural coastlines, is a measure of the rate of erosion, e.g. rate of slope retreat. For man-made structures the frequency

of repair, for example, reflects the rate at which the structure has deteriorated and the condition is a measure of the present serviceability.

Public safety

This factor assesses the potential threat of deterioration to public safety on the basis of severity of failure and public accessibility.

Recommended action

This system is currently being developed so that the scoring system can be calibrated against a number of case histories. The total score is used to prioritize a programme of remedial measures (Fig. 2)

Location

A classification scheme has been developed here using two case studies from the south Devon coast. Both localities form part of the itinerary for one of the University of Greenwich MSc Geomaterials Field Courses and have been developed from student exercises over several years.

Example of a natural coastal section

The coastal section area consists of fault-bounded blocks of Devonian limestones and shales, and Triassic sandstones and conglomerates, forming cliffs up to

Fig. 3. Coastal section.

GENERAL LOCATION AND REFERENCE DETAILS.
Footbridge over railway

1. _STRUCTURE_	LANDWARD PIER					SEAWARD PIER					RAIL WALL	PATHWAY	ROCK ARMOUR
1A FOOTBRIDGE													

2. _GEOLOGY_													

3. _MATERIAL OF CONSTRUCTION_	LANDWARD PIER					SEAWARD PIER					RAIL WALL	PATHWAY	ROCK ARMOUR
3A STONE/MASONRY					✓					✓	✓		✓
3B CONCRETE				✓					✓			✓	
3C BITUMOUS BOUND												✓	
3D TIMBER			✓					✓					
3E METAL		✓					✓						
3F BRICK	✓					✓							
3G FILL													
3H OTHER													

4. _AGE_	4	4	4	9	4	4	4	4	9	4	4	9	9

5. _ENVIRONMENT_	5	5	5	5	5	7	7	7	7	7	7	7	9

6. _COASTAL PROCESS_													

7. _FREQUENCY OF REPAIR_	1	2	2	2	1	1	2	2	2	1	2	5	1

8. _RATE OF DETERIORATION_													

9. _CONDITION_	2	9	6	4	1	2	10	7	5	1	2	3	3

10. _PUBLIC SAFETY_	3	7	5	5	3	3	7	5	5	3	3	4	3

TOTALS	15	27	22	25	14	17	30	25	28	16	18	28	25

11. RECOMMENDED ACTION													
REPAIR URGENTLY							✓						
REPAIR RECOMMENDED		✓		✓				✓	✓			✓	
MONITOR CONTINUOUSLY			✓										✓
INSPECT ANNUALLY											✓		
INSPECT AFTER 5 YRS	✓				✓	✓				✓			

Fig. 4. Example of a completed pro forma for a structure.

Fig. 5. Victorian footbridge, Dawlish.

100 m high (Fig. 3). It has been a popular amenity since Victorian times and there are several coastal defence structures of this age as well as a number of more recent slope remediation measures. The classification system provides a framework for the assessment of risk, both to the amenities and to the public, and the formulation of a remedial action plan.

Example of a man-made structure

Brunel's railway between Exeter and Teignmouth, east of Dawlish, runs along a low embankment on the old foreshore and is protected by both sea wall and rock armour. The completed pro forma (Fig. 4) illustrates the assessment of one particular cross-section which includes a Victorian footbridge over the railway (Fig. 5). The pro forma allows the evaluation of the relative durability of the various bridge and coastal rock armour materials for the effective design of future remedial measures.

Summary

The classification methods described here are a model of what can be used by relatively inexperienced engineers and provide a rapid and cost-effective means for initial assessment so that a maintenance strategy can be implemented. The general methodology has been used successfully, for example to survey marine concrete structures in the Middle East (Fookes *et al.* 1981) and in the UK for a Plymouth car park (Fookes *et al.* 1983a, b, 1984). Each site requires adaptation of the general scheme to suit local conditions and objectives.

References

BIENIAWSKI, I. T. 1974. Geomechanics classification of rock masses and its application to tunnelling. *Proceedings of the Third Congress of the International Society on Rock Mechanics*, Denver, **1**, 27–32.

FOOKES, P. G. 1997. Geology for engineers: the geological model, prediction and performance. *Quarterly Journal of Engineering Geology*, **3**, 294–424.

FOOKES, P. G., POLLOCK, O. J. & KAY, E. A. 1981. Middle East Concrete (2) – Rates of deterioration. *Concrete*, **15**(9), 12–19.

FOOKES, P. G., COMBERBACH, C. D. & CANN, J. 1983a. Field investigation of concrete structures in South-west England (1). *Concrete*, **17**(3), 54–56.

FOOKES, P. G., COMBERBACH, C. D. & CANN, J. 1983b. Field investigation of concrete structures in South-west England (2). *Concrete*, **17**(4), 60–65.

FOOKES, P. G., CANN. J. & COMBERBACH, C. D. 1984. Field investigation of concrete structures in South-west England (3). *Concrete*, **17**(4), 12–16.

Predicting natural cavities in chalk

C. N. Edmonds

Peter Brett Associates, Reading, Berkshire, UK

Introduction

Chalk is a soluble carbonate rock with extensive karst development. Natural cavity occurrence initially appears to be random. In an area where the degree of influence of all cavity formational factors is similar, but dissolution is focused on one set of joints rather than another, then solution feature occurrence is perhaps random. This might be termed the 'microscale' view, measured at a scale of metres. However, if the pattern of natural cavity occurrence is considered at a 'macroscale' level, say measured in hundreds of metres or kilometres, then spatial patterns emerge suggesting dissolution is not as random as it might first appear.

Spatial characteristics of natural cavity occurrence on the Chalk

In order to analyse the spatial characteristics of natural cavity occurrence it was first necessary to collect as many records of solution feature occurrence as possible from published and unpublished sources including local authorities, Construction Industry Research and Information Association (CIRIA), Transport Research Laboratory (TRL), Building Research Establishment (BRE), National House Building Council (NHBC), the Environment Agency, water companies, and site investigation reports. Particular emphasis was placed on fieldwork to record new features revealed in a wide range of engineering works, road and motorway construction, and on visits to large numbers of working/disused chalk quarries and aggregate workings. A database of 2226 natural cavities was compiled, composed mainly of solution pipes, sinkholes (dolines) and swallow holes.

The spatially related database (each cavity location being recorded by National Grid Reference) was carefully scrutinized to determine the conditions commonly associated with cavity occurrence (Edmonds 1987). The spatial analysis was undertaken with reference to the ideas previously put forward by Higginbottom (1979) on factors that appeared to influence natural cavity occurrence on the Chalk that were linked to subsidence risk levels.

The research found that for natural cavities to be formed, certain conditions needed to be present. Visual comparison of cavity occurrence versus 13 qualitative geological, hydrogeological and geomorphological factors (Table 1) was undertaken (Edmonds *et al.* 1987) for the natural cavities database. The following points serve to illustrate some of the main conclusions of the research.

(i) Ninety-seven per cent of natural cavities occur on the Upper Chalk, 2% on Middle Chalk and 1% on Lower Chalk.

(ii) Lower than average numbers of natural cavities are associated with the harder chalks of Lincolnshire and Yorkshire, and with the tectonically hardened chalks (e.g. The Hogs Back, Surrey).

(iii) Ninety-seven per cent of natural cavities are associated with the presence of a post-Cretaceous cover deposit (such as Palaeogene and Quaternary deposits). Only 3% of features are isolated from a cover deposit, but still generally occur within 200 m of a cover deposit margin.

Table 1. *Geological hydrogeological and geomorphological factors*

Geological factors
- Lithostratigraphic horizons of the Chalk
- Biostratigraphic horizons of the Chalk
- Tectonic structure
- Presence of post-Cretaceous cover deposits overlying the Chalk
- Lithology of post-Cretaceous cover deposits

Hydrogeological factors
- Hydrogeological characteristics of the Chalk
- Hydrogeological characteristics of post-Cretaceous cover deposits
- Water table level in relation to the Chalk surface and cover deposit interface levels
- Effects of topographic relief upon surface water drainage and subsurface groundwater infiltration

Geomorphological factors
- Locations of former surface water drainage paths
- Effects of marine planation
- Effects of glaciation
- Effects of periglaciation

From: GRIFFITHS, J. S. (ed.) *Land Surface Evaluation for Engineering Practice*. Geological Society, London, Engineering Geology Special Publications, **18**, 29–38. 0267-9914/01/$15.00 © The Geological Society of London 2001.

(iv) Natural cavities are recorded below both cohesive and granular cover deposits up to 45 m in thickness.

(v) Ninety-two per cent of natural cavities are associated with surface water drainage and groundwater infiltration flows directed onto/into the Chalk from overlying Palaeogene and Quaternary deposits.

(vi) The potential for natural cavity occurrence increases with the increasing tendency of topographic relief and surface water drainage/groundwater infiltration conditions to concentrate water flows into the ground. The land surface may be subdivided into water flow 'concentrators'. Fifty-three per cent of natural cavities are associated with concave surface channels, 28% are associated with concave/convex side slopes and 19% are associated with other land surface forms.

The role of former drainage paths, especially the proto-Thames and proto-Solent river systems, is thought to have been influential in the formation of solution features in the past. It is notable that higher numbers of natural cavities are recorded within the Chilterns Hills and the Dorset and Purbeck Downs, where these extensive river systems were formerly active. It appears that they must have caused large volumes of water to percolate down into the Chalk resulting in much dissolution activity.

Other influences, such as marine planation and glacial erosion, can be both destroyers and creators of natural cavity-forming conditions. If marine or glacial erosion results in removal of a cover deposit and the chalk surface zone containing natural cavities, then clearly natural cavities are destroyed and the numbers of natural cavities greatly reduced. Conversely, if, following the erosion, a new cover deposit is laid down and the passage of concentrated water flows down into or onto the Chalk is resumed, then a new phase of active natural cavity formation commences. This is a feature of natural cavity occurrence noted in East Anglia associated with the glacial cover deposits. However, in the Lincolnshire and Yorkshire Chalk areas adjoining the present coastline, the combination of marine and glacial erosion has etched

deeply into the Chalk such that it is fully saturated or water flows in the Chalk are upwards (artesian) below the glacial cover deposits. In these circumstances there are little or no opportunities for the downward flow of water into the Chalk from the cover deposit, hence no natural cavities are recorded in the unfavourable circumstances.

In extraglacial areas it is considered that periglacial weathering has been generally beneficial to natural cavity formation. In particular, it is thought that the downward percolation of carbon-dioxide-rich, cold groundwater, released by the melting of ground ice formed within cover deposits overlying the Chalk, is responsible for the formation and reactivation of many natural cavities. It seems likely that this would have been the main source of water infiltration down into the Chalk that was responsible for natural cavity creation below cohesive cover deposits such as Reading Beds clays and Clay with Flints (Plateau Drift).

Development of natural cavity occurrence (hazard) mapping techniques

The qualitative spatial analysis described above established substantial evidence for causal relationships between natural cavity occurrence and certain controlling factors. Following this, the controlling factors were ranked into major, moderate and minor influential factors as shown in Table 2. These are the factors that were consistently found to have the most pronounced and specific influence on cavity formation.

The next step in the development of the hazard mapping technique was to express the qualitative influential factors in a quantitative way. It was decided to quantify major factors using a scale of 0 to 20, moderate factors using a scale of 0 to 10, and minor factors using a scale of 0 to 5. The scale ranges were intended to reflect the relative degree of influence of the different factors. The higher the number, the higher the potential for natural cavity occurrence and associated instability hazard.

The specific numerical values for each of the influential factors are shown in Tables 3 to 8 for the varying

Table 2. *Major, moderate and minor influential factors*

Major influential factors
- Chalk lithostratigraphy
- Post-Cretaceous cover deposits

Moderate influential factors
- Water table level
- Topographical relief and surface water drainage/subsurface groundwater infiltration
- Former surface water drainage path

Minor influential factor
- Glaciation

Table 3. *Chalk lithostratigraphic factor (G_1)*

Choose one	Value
Upper chalk	20
Middle chalk	2
Lower chalk	1

Guidance note
1. Choose the appropriate Chalk lithostratigraphic unit that underlies the surface below the area of interest and carry forward the numerical value for insertion into the formula in Table 9.

Table 4. *Post-Cretaceous cover deposit factor (G_2) (note: $G_2 = Gc + Gr + Gf$)*

Gc		Gr		Gf	
Choose one	*Value*	*Choose one*	*Value*	*Choose one*	*Value*
(a) Tertiary cover deposit present with or without superficial Quaternary cover	14	*Tertiary cover deposit present:* (a) Reading Beds – Woolwich Beds (all regions)	20	(a) Reading Beds feathering margin	3
(b) Quaternary cover deposit only present	6	(b) Thanet Beds (WND, END regions)	15	(b) Thanet Beds feathering margin	4
(c) No cover deposit present, but within 200 m of a Tertiary or Quaternary cover deposit margin	2	(c) Crag (EA region)	8	(c) Crag feathering margin	2
(d) No cover deposit present and >200m from a Tertiary or Quaternary cover deposit margin	1	(d) Thicker Tertiary sequences involving Thanet Beds overlain by Woolwich–Reading Beds, and Blackheath Beds and disturbed Blackheath beds (WND region)	4	(d) Tertiary margins where thick Tertiary sequences occur (WND, DPD regions)	1
		(e) Thicker Tertiary sequences involving Reading Beds overlain and overstepped by London Clay and Bagshot Beds (DPD region)	2	(e) Quaternary deposit feathering margin	2
		Quaternary cover deposit present: (a) Proto-Thames and Proto-Solent terrace gravels (CH and DPD regions)	20	(f) Exception to (e) for alluvial deposits where seasonal drainage is not directed across Chalk (Category 2) of topographic relief and surface drainage/subsurface infiltration factor (H_2)	0
		(b) Alluvial deposits (all regions) (only applicable to seasonal drainage directed across Chalk (Category 2) of topographic relief and surface drainage/subsurface infiltration factor (H_2)	13	(g) No cover deposit (topsoil only)	0
		(c) Alluvial deposits (all regions) (any topographic relief and surface drainage/subsurface infiltration condition except above)	1	(h) No feathering margins	0
		(d) Glacial deposits (CH, EA, LY regions)	12		
		(e) Low level fluvial terrace gravels or valley gravel (all regions)	6		
		(f) High level fluvial terrace gravels or plateau gravel (all regions)	6		
		(g) Clay-with-flints (all regions)	3		
		(h) Brickearth (all regions)	1		
		(i) Solifluction deposits (all regions)	1		
		(j) No cover (topsoil only) (all regions)	0		

Guidance notes
1. The value for factor G_2 is calculated by summing together the values of the subfactors Gc, Gr and Gf.
2. For reference to the chalkland regions see Fig. 1.
3. When determining Gr, the presence of a Tertiary deposit, even if overlain by Quaternary deposit, takes precedence when choosing the appropriate Gr value.
4. For the area of interest, choose the one appropriate set of conditions from each of Gc, Gr and Gf in turn to derive the G_2 value. Carry forward the G_2 value for insertion into the formula in Table 9.

circumstances as they apply to the Chalk. A number of the tables make reference to particular conditions that apply to certain chalkland regions. To assist with interpreting appropriate values for the factors, Figure 1 shows the chalkland regions as used by the research for subdividing the Chalk outcrop and subcrop in southern and eastern England.

The following account provides a summary of the procedures for applying the hazard mapping techniques to a particular area of interest. In the first instance the

Table 5. *Water table level factor (H₁)*

Choose one	Value
(a) No cover deposit present, water table level below Chalk surface level	10
(b) No cover deposit present, water table close to or at Chalk surface level	1
(c) Cover deposit present, water table below Chalk/ cover deposit interface	10
(d) Cover deposit present, seasonal water table level fluctuation causes water table level to rise above Chalk/cover deposit interface in wet season	5
(e) Cover deposit present, water table level normally at or above Chalk/cover deposit interface	1
(f) Same as (e), but if artificial groundwater lowering to take place	3
(g) Artesian groundwater conditions present in cover deposit overlying Chalk	0

Guidance notes
1. The conditions described in (d) to (g) only apply where the cover deposit and the Chalk are in hydraulic continuity with one another. They do not apply to situations where a separate perched water level exists within the cover deposit that is unrelated to the water table level within the Chalk below.
2. Carry forward the H_1 value for insertion into the formula in Table 9.

information to derive appropriate values for factors G_1, G_2 and H_1 (see Tables 3–5) may be obtained by referring to published geological and hydrogeological maps for the area of interest. Applying factor H_2 to the area of interest requires the area to be geomorphologically mapped to subdivide the land surface into terrain units that reflect the surface drainage/subsurface infiltration characteristics, the so-called water flow 'concentrators', referred to above. In order to visualize these characteristics, Figure 2 shows a series of block models to demonstrate common water flow scenarios found in the Chalk and Figure 3 demonstrates the principle of subdividing the landscape into mappable water flow concentrator units. This work, carried out by walkover survey supported by reference to topographic maps and aerial photograph interpretation, forms a precursor to the hazard map finally produced. Information on former drainage paths (factor GM_1) can be obtained from published literature (e.g. Jones 1981) and the details for GM_2 can be derived from published geological maps.

The numerical values quoted in the tables for the spatial controlling factors were only arrived at after much detailed analysis and experimentation. Iterative application, assessment and reassessment of the results, when the factors were applied to the recorded natural cavities database, was executed. Wherever possible,

Table 6. *Topographic relief and surface drainage/subsurface infiltration factor (H₂)*

Choose one category

Category 1 Seasonal/permanent surface drainage and subsurface infiltration directed onto/ into Chalk from cover deposit		Category 2 Seasonal surface drainage and subsurface infiltration directed towards/onto cover deposit from Chalk		Category 3 Seasonal/permanent surface drainage and subsurface infiltration directed across Chalk (may be covered by topsoil and/or solifluction deposits or alluvium in floors of seasonal/permanent streams)	
Choose terrain unit	Value	Choose terrain unit	Value	Choose terrain unit	Value
Terrain Unit 1	6	Terrain Unit 1	1	Terrain Unit 1	1
Terrain Unit 2	10	Terrain Unit 2	2	Terrain Unit 2	3
Terrain Unit 3	4	Terrain Unit 3	1	Terrain Unit 3	1

Guidance notes
1. For the area of interest first consider the form of topographic relief and surface/subsurface water movement conditions that would apply to the area. This should be carried out with reference to Figure 2 to derive the category in the above table that most closely fits the situation. Support information will also need to be taken from the published geology, hydrogeology and topographic survey maps. Stereoscopic viewing of aerial photographs can also be helpful.
2. The next step is to subdivide the area of interest into terrain units which determine the degree of surface/subsurface water concentration. This exercise is carried out by performing standard geomorphological mapping of the area recording the concave and convex surfaces. This is best done by field survey combined with the use of aerial photograph interpretation. Finally a map of the terrain units is created for the area of interest, as illustrated by Figure 3. The recording of slope angles within terrain units is not particularly important for this exercise, the landform type is more important to establish. However, for guidance Terrain Unit 3 will have slope angles of zero to 1° or 2°, while Terrain Unit 2 often has slope angles in the range of 2° to 5° sometimes up to 10°.
3. Carry forward the H_2 values applicable and insert them, in turn, into the formula in Table 9 to calculate the SHR_N values.

Table 7. *Former surface water drainage path factor (GM₁)*

Choose one	Value
(a) Proto-Solent corridor (DPD region)	10
(b) Proto-Thames corridor (CH region) but not applicable to following:	10
(i) where glacial deposits directly overlie Chalk or,	0
(ii) where surface drainage/ subsurface infiltration is directed off the Chalk towards/onto a Tertiary cover	0
(c) Outside above proto-river corridors	0

Guidance notes
1. To check the geographic location and extent of the former proto-Thames and proto-Solent drainage path corridors it is advised that reference is made to appropriate published texts, e.g. D. K. C. Jones, *Southeast and Southern England*, University Paperbacks, Methuen, 1981.
2. For reference to the chalkland regions see Figure 1.
3. When $H_1 = 1$ factor GM_1 is not applicable hence $GM_1 = 0$ (DPD, CH regions).
4. If the area of interest lies within a former drainage path corridor then choose the appropriate situation to derive the factor value. Transfer the value for insertion into the formula in Table 9.

Table 8. *Glaciation factor (GM₂)*

Choose one	Value
(a) Glacial deposits directly overlie the Chalk (CH, EA, L, Y regions)	5
(b) Glacial deposits overlie Crag upon Chalk (EA region)	3
(c) Glacial deposits overlie Reading Beds and/or proto-Thames terrace gravels upon Chalk (CH region)	0
(d) Glacial deposits absent (CH, EA, L, Y regions)	0
(e) Extra-glacial areas	0

Guidance notes
1. To check the geographic location and extent of the formerly glaciated areas of the Chalk it is advised that reference is made to appropriate published texts, e.g. D. K. C. Jones, *Southeast and Southern England*, University Paperbacks, Methuen, 1981.
2. For reference to the chalkland regions see Figure 1.
3. If the area of interest lies within a formerly glaciated area then choose the appropriate situation to derive the factor value. Transfer the value for insertion into the formula in Table 9.

derived values were based closely upon the proportional influence shown by the recorded cavities database in response to a particular spatial factor. However, sometimes when derived values were reapplied to the database the potential for natural cavity occurrence (hazard)

was over- or underestimated, hence it was necessary to adjust certain values to overcome such problems. Eventually, utilizing the results of the natural cavity occurrence statistics from the research database, the factors were suitably numerically weighted against the chosen simple arithmetic scales. Finally, the derived scale value for each factor is entered into the mathematical formula shown in Table 9. In recognition that the occurrence of a natural cavity gives rise to the potential for land instability, the formula is utilized to derive a subsidence hazard rating value, SHR_N. Once the numerical value of SHR_N has been calculated it is compared with Table 10 to obtain the final subsidence hazard classification.

By this methodology the range of possible subsidence hazard classifications that apply to the area of interest are calculated. The classifications are then attached to each of the spatial landscape units to which they apply in order to produce the finished subsidence hazard map. This is further explained with reference to a case study where the techniques have been used to generate a subsidence hazard map for an area of northwest Reading, Berkshire (Edmonds 2001).

Limitations

It was recognized at an early stage that the database would show some bias towards urban areas where features had been discovered as a result of man's activities. However, despite this it became apparent that in some areas features were plentiful while in others they were absent given similar opportunities for discovery by excavation. This was also reinforced by the field evidence in non-urban areas. It was determined that the pattern of solution feature occurrence is not an artefact of man's activities and nor is it of anthropogenic origin (Edmonds 1984).

Applications

Since the completion of the academic research, the natural cavities database has been greatly enlarged. The database forms the core of a national natural cavities database prepared for the Department of the Environment (now the Department of the Environment, Transport and the Regions), the results of which were published by Applied Geology Limited in 1994.

It has been found by hazard mapping that, in general, the higher the hazard rating the greater the statistical chance of natural cavities occurring, thereby increasing the potential for land instability as well (see Table 11). The technique has been applied to more than 200 sites underlain by Chalk. Consequently the results of the work have been subjected to academic and

Key:-

a   Yorkshire (Y)
b   Lincolnshire (L)
c   East Anglia (EA)
d   Chiltern Hills (CH)
e   Berkshire & Marlborough Downs (BMD)
f   Salisbury Plain (SP)
g   Hampshire Downs (HD)
h   West North Downs (WND)
i   East North Downs (END)
j   Thanet (T)
k   South Downs (SD)
l   Littlehampton (Li)
m   Portsdown (P)
n   Isle of Wight (IOW)
o   Cranborne Chase (CC)
p   Dorsetand Purbeck Downs (DPD)
q   Devon-Dorset Outliers (DDO)

Fig. 1. Plan of chalkland regions.

commercial scrutiny. The hazard mapping technique has been found to provide a generally reliable way of expressing cavity occurrence and subsidence potential. The hazard mapping approach and its uses in investigating sites underlain by Chalk is outlined in Kirkwood & Edmonds (1989).

The technique has been found to be useful in providing prior warning of potential subsidence problems and related effects in relation to new road construction and the development of housing, commercial and retail premises. On a number of occasions the method has been used to interpret and understand patterns of ground subsidence behaviour affecting exist-

ing development, as an aid to determining remedial measures. Currently, the technique is being adapted for use as a mapping tool to recognize the potential for contamination susceptibility of the Chalk aquifer due to karstic landforms.

Previous research

Research of karstic spatial patterns is not new: many accounts have been published from at least the 1960s onwards. For the Chalk, some of the earliest research

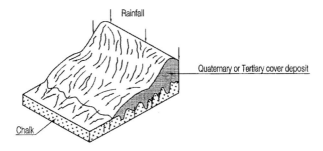

Category 1.

Surface drainage and groundwater infiltration originates upon Tertiary/Quaternary cover deposits, then flows onto and percolates down into chalk below.

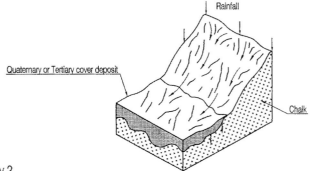

Category 2.

Surface drainage and groundwater infiltration originates upon chalk, then flows down onto and into Tertiary/Quaternary deposits at foot of slope.

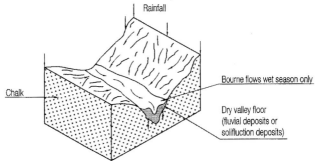

Category 3.

Surface drainage and groundwater infiltration originates upon chalk, then is directed down slope and concentrated in chalk valley floor with surface (bourne)flows along and surface percolation into the fluvial/solifluction deposits in filling the valley floor. Category normally applicable to upland chalk areas.

Fig. 2. Regional topographic relief and surface drainage/subsurface infiltration models.

in solution feature patterns was undertaken by Prince (1961, 1962, 1964) in East Anglia and later by Sperling *et al.* (1977, 1979) and Prince (1979) for the many sinkholes found on the Dorset heathlands. Elsewhere, examples of statistical studies include McConnell & Horn (1972), Williams (1972), Troester *et al.* (1984) and Thorp & Brook (1984). Other studies have analysed spatial patterns by visual comparison, relying strongly on the use of remote sensing and/or geophysics (e.g. Black 1984; Littlefield *et al.* 1984; Stewart & Wood 1984).

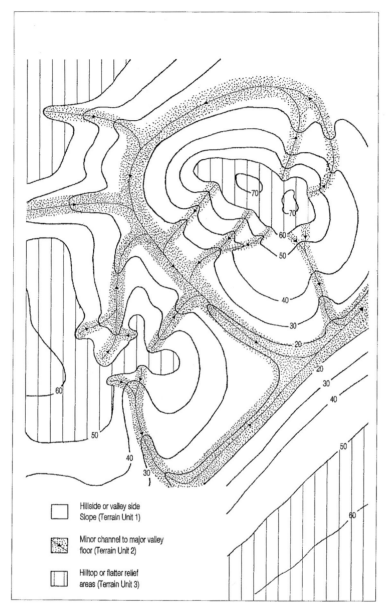

Fig. 3. Schematic illustration of simplified topographic relief and surface drainage/subsurface infiltration terrain unit mapping.

Further studies have considered combinations of possible controlling factors such as stratigraphy, lithology, structure, hydrogeology, topography and geomorphology using multilayered datasets for visual comparison (e.g. Benson & la Fountain 1984; Dalgleish & Alexander 1984; Day 1984; Fischer & Greene 1984; Hubbard 1984; Ogden 1984). Once spatial pattern controlling factors were identified they were combined qualitatively or semi-quantitatively to produce simple subsidence hazard predictive models.

Table 9. *Subsidence hazard mapping formula*

Formula: $SHR_N = (G_1 + G_2 + H_1 + GM_1 + GM_2)H_2$

Factor identities for the formula are shown below:

SHR_N = subsidence hazard rating for natural cavities
  G_1 = chalk lithostratigraphic factor
  G_2 = post-Cretaceous cover deposit factor
  H_1 = water table level factor
  H_2 = topographic relief and surface drainage/subsurface
     infiltration factor
  GM_1 = former drainage path factor
  GM_2 = glaciation factor

Guidance note
1. Substitute factor values from Tables 3 to 8 into formula and calculate SHR_N value.
2. Compare SHR_N value with Table 10.

Table 10. *Subsidence hazard classification*

SHR_N	Subsidence hazard category
<55	No anticipated subsidence hazard
55–89	Very low subsidence hazard
90–136	Low subsidence hazard
137–200	Moderately low subsidence hazard
201–300	Moderate subsidence hazard
301–400	Moderately high subsidence hazard
401–600	High subsidence hazard
>600	Very high subsidence hazard

Guidance notes
1. Use SHR_N value to derive subsidence hazard category.
2. Reapply the subsidence hazard category to the terrain unit for which all derivative factors were originally obtained.
3. Repeat the procedure for all other terrain units within area of interest to create a subsidence hazard map of the area.

A series of papers by Dougherty (1989), Fischer & Lechner (1989) and Kochanov (1989) illustrate how, in the USA, natural cavity occurrence mapping and hazard zoning is being used for development land management by statutory authorities in Pennsylvania and New Jersey.

Further examples of modelling technique approaches for deriving land instability hazard ratings for areas prone to mining subsidence, landsliding and limestone subsidence are presented by papers in Culshaw *et al.* (1987). Similar ideas and approaches to understanding subsidence behaviour and risk on the Chalk have been published more recently by Rigby-Jones *et al.* (1993) and for a variety of geohazards in Maund & Eddleston (1998). The development, application and usage of hazard mapping methods by engineering geomorphologists is becoming more common for evaluating a wide range of geohazards.

Table 11. *Statistical relationship of cavity occurrence with hazard category*

Subsidence hazard category	Number of solution features	Percentage of solution features
No anticipated hazard	2	0.1
Very low hazard	10	0.5
Low hazard	62	3
Moderately low hazard	73	3.6
Moderate hazard	147	7.2
Moderately high hazard	244	11.9
High hazard	466	22.8
Very high hazard	1042	50.9
Totals	2046	100

Guidance notes
1. The table shows the increasing numbers of solution features (natural cavities) that are found within each of the ascending hazard categories when the subsidence hazard mapping technique is applied to the research database (2226 natural cavity records).
2. The total number of solution features is 2046, rather than 2226, because 180 features could not be located sufficiently accurately to determine the hazard category that was applicable.

References

APPLIED GEOLOGY LIMITED. 1994. *Review of Instability due to Natural Underground Cavities in Great Britain.* Vols 1.1 to 1.10, 2.1 to 2.3 and Summary Report for the Department of the Environment.

BENSON, R. C. & LA FOUNTAIN, L. J. 1984. Evaluation of subsidence or collapse potential due to subsurface cavities. *In*: BECK, B. F. (ed.) *Sinkholes: their Geology, Engineering and Environmental Impact.* Proceedings of the First Multidisciplinary Conference on Sinkholes, Florida, Balkema Press, 201–215.

BLACK, T. J. 1984. Tectonics and geology in karst development of Northern Lower Michigan. *In*: BECK, B. F. (ed.) *Sinkholes: their Geology, Engineering and Environmental Impact.* Proceedings of the First Multidisciplinary Conference on Sinkholes, Florida, Balkema Press, 87–91.

CULSHAW, M. G., BELL, F. G., CRIPPS, J. C. & O'HARA, M. 1987. *Planning and Engineering Geology.* Proceedings of the 22nd Annual Conference of the Engineering Group of The Geological Society. Engineering Geology Special Publications, **4**, The Geological Society, London.

DALGLEISH, J & ALEXANDER, E. C. 1984. Sinkhole distribution in Winona County, Minnesota. *In*: BECK, B. F. (ed.) *Sinkholes: their Geology, Engineering and Environmental Impact.* Proceedings of the First Multidisciplinary Conference on Sinkholes, Florida, Balkema Press, 79–85.

DAY, M. 1984. Predicting the location of surface collapse within karst depressions: A Jamaican example. *In*: BECK, B. F. (ed.) *Sinkholes: their Geology, Engineering and Environmental Impact.* Proceedings of the First Multidisciplinary Conference on Sinkholes, Florida, Balkema Press, 147–151.

DOUGHERTY, P. H. 1989. Land use regulations in the Lehigh Valley: Zoning and subdivision ordinances in an environmentally sensitive karst region. *In*: BECK, B. F. (ed.)

Engineering and Environmental Impacts of Sinkholes and Karst. Proceedings of the Third Multidisciplinary Conference on Sinkholes, St. Petersburg Beach, Florida, Balkema Press, 341–348.

EDMONDS, C. N. 1984. Reply to Discussion 'Towards the prediction of subsidence risk upon the Chalk outcrop' by, C. N. Edmonds. *Quarterly Journal of Engineering Geology*, **17**, 167–168.

EDMONDS, C. N. 1987. *The engineering geomorphology of karst development and the prediction of subsidence risk upon the Chalk outcrop in England*. PhD Thesis, University of London.

EDMONDS, C. N. 2001. Subsidence hazards in Berkshire in areas underlain by Chalk karst. *This volume*.

EDMONDS, C. N. , GREEN, C. P. & HIGGINBOTTOM, I. E. 1987. Subsidence hazard prediction for limestone terrains, as applied to the English Cretaceous Chalk. *In*: CULSHAW, M. G., BELL, F. G., CRIPPS, J. C. & O'HARA, M. (eds) *Planning and Engineering Geology*. Proceedings of the 22nd Annual Conference of the Engineering Group of The Geological Society. Engineering Geology Special Publications, **4**, The Geological Society, London, 283–293.

FISCHER, J. A. & GREENE, R. W. 1984. New Jersey sinkholes: Distribution, formation, effects – Geotechnical Engineering. *In*: BECK, B. F. (ed.) *Sinkholes: their Geology, Engineering and Environmental Impact*. Proceedings of the First Multidisciplinary Conference on Sinkholes, Florida, Balkema Press, 159–165.

FISCHER, J. A. & LECHNER, H. 1989. A karst ordinance – Clinton Township, New Jersey. *In*: BECK, B. F. (ed.) *Engineering and Environmental Impacts of Sinkholes and Karst*. Proceedings of the Third Multidisciplinary Conference on Sinkholes, St. Petersburg Beach, Florida, Balkema Press, 357–361.

HIGGINBOTTOM, I. E. 1979. *Suggested distribution of chalk pipes and swallow holes with a tentative indication of relative subsidence risk levels*. Unpublished presentation to the Geological Society of London.

HUBBARD, D. A. 1984. Sinkhole distribution in the central and northern Valley and Ridge province, Virginia. *In*: BECK, B. F. (ed.) *Sinkholes: their Geology, Engineering and Environmental Impact*. Proceedings of the First Multidisciplinary Conference on Sinkholes, Florida, Balkema Press, 75–78.

JONES, D. K. C. 1981. *Southeast and Southern England*. University Paperbacks, Methuen.

KIRKWOOD, J. P. & EDMONDS, C. N. 1989. Ground subsidence movements upon the Cretaceous chalk outcrop in England – Sinkhole problems and engineering solutions. *In*: BECK, B. F. (ed.) *Engineering and Environmental Impacts of Sinkholes and Karst*. Proceedings of the Third Multidisciplinary Conference on Sinkholes, Florida, Balkema Press, 247–255.

KOCHANOV, W. E. 1989. Karst mapping and applications to regional land management practices in the Commonwealth of Pennsylvania. *In*: BECK, B. F. (ed.) *Engineering and Environmental Impacts of Sinkholes and Karst*. Proceedings of the Third Multidisciplinary Conference on Sinkholes, Florida, Balkema Press, 363–368.

LITTLEFIELD, J. R., CULBRETH, M. A., UPCHURCH, S. B. & STEWART, M. T. 1984. Relationship of modern sinkhole development to large scale photolinear features. *In*: BECK,

B. F. (ed.) *Sinkholes: their Geology, Engineering and Environmental Impact*. Proceedings of the First Multidisciplinary Conference on Sinkholes, Florida, Balkema Press, 189–195.

MAUND, J. G. & EDDLESTON, M. 1998. *Geohazards in Engineering Geology*. The Geological Society, London, Engineering Geology Special Publications, **15**.

MCCONNELL, H. & HORN, J. M. 1972. Probabilities of surface karst. *In*: CHORLEY, R. J. (ed.) *Spatial Analysis in Geomorphology, Part II – Point Systems*. Methuen, London, 111–133.

OGDEN, A. E. 1984. Methods for describing and predicting the occurrence of sinkholes. *In*: BECK, B. F. (ed.) *Sinkholes: their Geology, Engineering and Environmental Impact*. Proceedings of the First Multidisciplinary Conference on Sinkholes, Florida, Balkema Press, 177–182.

PRINCE, H. C. 1961. Some reflections on the origin of hollows in Norfolk compared with those in the Paris region. *Revue de Geomorphologie Dynamique*, **12**, 110–117.

PRINCE, H. C. 1962. Pits and ponds in Norfolk. *Erdkunde*, **16**, 10–31.

PRINCE, H. C. 1964. The origin of pits and depressions in Norfolk. *Geography*, **49**, 15–32.

PRINCE, H. C. 1979. Discussion – Marl pits or dolines of the Dorset Chalklands. *Transactions of the Institute of British Geographers, New Series*, **4**, 116–117.

RIGBY-JONES, J., CLAYTON, C. R. I. & MATTHEWS, M. C. 1993. Dissolution features in the chalk: from hazard to risk. Paper 7. *In*: INSTITUTION OF CIVIL ENGINEERS (eds) *Risk and Reliability in Ground Engineering*. Thomas Telford, London, 87–99.

SPERLING, C. H. B., GOUDIE, A. S., STODDART, D. R. & POOLE, G. G. 1977. Dolines of the Dorset Chalklands and other areas of Southern Britain. *Transactions of the Institute of British Geographers, New Series*, **2**, 205–223.

SPERLING, C. H. B., GOUDIE, A. S., STODDART, D. R. & POOLE, G. G. 1979. Discussion – Origin of the Dorset Dolines. *Transactions of the Institute of British Geographers, New Series*, **4**, 121–124.

STEWART, M. & WOOD, J. 1984. Geophysical characteristics of fracture traces in the carbonate Floridan Aquifer. *In*: BECK, B. F. (ed.) *Sinkholes: their Geology, Engineering and Environmental Impact*. Proceedings of the First Multidisciplinary Conference on Sinkholes, Florida, Balkema Press, 225–229.

THORP, M. J. W. & BROOK, G. A. 1984. Application of double Fourier series analysis to ground subsidence susceptibility mapping in covered karst terrain. *In*: BECK, B. F. (ed.) *Sinkholes: their Geology, Engineering and Environmental Impact*. Proceedings of the First Multidisciplinary Conference on Sinkholes, Florida, Balkema Press, 197–200.

TROESTER, J. W., WHITE, E. L. & WHITE, W. B. 1984. A comparison of sinkhole depth frequency distributions in temperate and tropic karst regions. *In*: BECK, B. F. (ed.) *Sinkholes: their Geology, Engineering and Environmental Impact*. Proceedings of the First Multidisciplinary Conference on Sinkholes, Florida, Balkema Press, 65–73.

WILLIAMS, P. W. 1972. The analysis of spatial characteristics of karst terrains. *In*: CHORLEY, R. J. (ed.) *Spatial Analysis in Geomorphology, Part II – Point Systems*. Methuen, London, 135–163.

Engineering geological mapping

J. S. Griffiths

Department of Geological Sciences, University of Plymouth, Devon, UK

Introduction

The development, scope and examples of best practice in the preparation of geological maps and plans for engineering practice are discussed in detail in Dearman (1991), and this remains the definitive British text on the subject. In the UK engineering geological maps were recognized as forming a significant component of site investigations with the publication of the Geological Society Working Party Report on maps and plans (Anon. 1972). In many parts of the world the practice is also long established (e.g. Peter 1966; Popov *et al.* 1950), and in 1976 UNESCO produced a guide for the preparation of engineering geological maps (Anon. 1976). Many examples of international studies were reported at the 1979 IAEG Newcastle Symposium on Engineering Geological Mapping. In contrast, in the United Kingdom, BS5930, the *Code of Practice for Site Investigations*, which was first published in 1981 and has guided the scope and content of site investigations in the UK for the past two decades, makes scant reference to geological or engineering geological mapping (Griffiths & Marsh 1986). As noted in Griffiths & Edwards (2001), this situation has not been corrected in the revision of BS5930 published in 1999. Fookes (1997), in the first Glossop lecture, states 'that engineering geological mapping, even sketch mapping, is particularly under used in British practice', and this is despite its longstanding and successful track record (Dearman & Fookes 1974). As with many facets of land surface evaluation, this situation needs to be addressed.

The distinction between engineering geological maps and plans

A critical aspect of all maps is the scale, and for engineering geological mapping the system contained in the UNESCO system is the most appropriate (Anon. 1976): large-scale maps are 1:10 000 or larger: medium-scale are between 1:10 000 and 1:100 000; and small-scale are 1:100 000 or smaller. Most engineering studies will require work to be carried out on large-scale maps or plans, although for regional studies and planning pur-

poses, medium- or, more rarely, small-scale maps can be appropriate. In detailed engineering geological studies, plans and sections at scales as large as 1:500 or even 1:100 are likely to be required.

Components of an engineering geological map

Fookes (1969) states that the aim of engineering geological mapping should be to produce a map on which the mapped units are defined by engineering properties or behaviour, and the limits of the units are determined by changes in the physical and mechanical properties of the materials. Clearly on such maps the boundaries of the mapped units may not correlate or coincide with the underlying geological structure or the lithostratigraphic units depicted on normal geological maps.

The information that has to be recorded on the engineering geological map was laid down in the Geological Society Working Party Report (Anon. 1972) and, with certain additions, these still provide an excellent checklist. An updated version of this list is presented as Table 1.

In addition to the observed mapped data, information on the location of previous site investigations should be noted including the sites of boreholes, trial pits and geophysical surveys. Similarly, notes should be made on all mines and quarries, including whether active or abandoned, dates of working, materials extracted and whether or not mine plans are available.

Whilst this list provides a broad coverage, the requirements for an individual engineering geological map can be tailored to suit the specific issues to be investigated. For example, in an area of volcanic risk there is likely to be more emphasis on the nature, frequency of occurrence and runout distances of lava flows, pyroclastic flows, lahars, ash clouds, lateral blasts and toxic gases. Similar detailed forms of investigation will be appropriate for different types of geohazard evaluation. It is also possible that more than one map will be required.

Dearman (1991) makes reference to the value of engineering geological zoning. This identifies areas on the map that have approximately homogenous engineering geological conditions. Such zones would normally be derived from the factual data compiled on the base

From: GRIFFITHS, J. S. (ed.) *Land Surface Evaluation for Engineering Practice*. Geological Society, London, Engineering Geology Special Publications, **18**, 39–42. 0267-9914/01/$15.00 © The Geological Society of London 2001.

Table 1. *Information to be recorded on an engineering geological map*

Geological data
- Mappable units (on the basis of descriptive engineering geological terms)
- Geological boundaries (with accuracy indicated)
- Description of soils and rocks (using engineering codes of practice, e.g. BS5930)
- Description of exposures (cross-referenced to field notebooks)
- Description of state of weathering and alteration (notes depth and degree of weathering)
- Description of discontinuities (as much detail as possible on the nature, frequency, inclination and orientation of all joints, bedding, cleavage, etc)
- Subsurface conditions (provision of subsurface information if possible, e.g rockhead isopachytes)

Hydrogeological data
- Availability of information (reference to existing maps, well logs, abstraction data)
- General hydrogeological conditions (notes on: groundwater flow lines; piezometric conditions; water quality; artesian conditions; potability)
- Hydrogeological properties of rocks and soils (aquifers, aquicludes and aquitards; permeabilities; perched water tables)
- Springs and seepages (flows to be quantified wherever possible)

Geomorphological data (see section on geomorphological mapping)
- General geomorphological features (ground morphology, landforms, processes)
- Ground movement features (landslides, subsidence, solifluction lobes; cambering)

Geohazards
- Mass movement (extent and nature of landslides, type and frequency of landsliding, possible estimates of runout hazard)
- Flooding (areas at risk, flood magnitude and frequency, coastal or river flooding)
- Coastal erosion (cliff form, rate of coastal retreat, coastal processes, types of coastal protection)
- Seismicity (seismic hazard assessment)
- Vulcanicity (volcanic hazard assessment)

map and therefore should not form part of the original map but can be produced as an overlay. Zoning maps can be particularly effective in geohazard studies where the magnitude of the hazard can often be represented by an interpretative map with an ordinal scaling for the degree of hazard.

Engineering geological plans

When engineering geological data are represented at scales larger than 1:10 000 then the material is shown on a plan. Engineering geological plans are made for specific civil engineering purposes and will normally be based on site plans made specifically for the location. As defined in Anon. (1972), the engineering geological plan is taken to include both maps and other methods of displaying field data such as cross-sections, gate diagrams and exposure logs. In these circumstances the mapping techniques used in smaller-scale mapping might have to be supplemented by detailed surveying and logging procedures to ensure the data are recorded with the necessary level of accuracy. Dearman (1991) divides plans into two types.

1. The pre-construction or site investigation plan. This is prepared during the early stages of an investigation to order to allow the ground investigations to be planned and engineering problems to be anticipated. These plans would normally be at 1:5000 for general purposes although may increase to 1:1000 for specific studies, for example of dam sites or tunnel portals (see Griffiths 2001).

2. Construction or foundation plans. These are produced during construction when exposures can be logged as foundations are excavated. These provide a record of the actual ground conditions and allow the geological ground model to be refined. The scales for plans may be as large as 1:100.

Given the potential for construction claims associated with unforeseen ground conditions, accurate engineering geological plans clearly can play a significant role in supporting or rejecting such petitions.

Data collection

Primary mapping for engineering geology follows the same basic rules and uses the same techniques established for conventional geological mapping (see Barnes 1997). All graduates in geology in the UK are taught these skills, thus engineering geological mapping should not prove to be a major problem to them. The initial decisions to be made when undertaking engineering geological mapping are to identify the types of data that are to be collected to meet the survey requirements. at what scale will mapping be carried out, and what methods are to be used for data collection. The required end-product must guide these decisions, i.e. what is the map to be used for?

In most engineering situations there will be three phases to the work: desk studies, including aerial photograph interpretation, where all existing data are compiled; primary mapping in the field; interpretation and preparation of the final maps. In most instances there will be a requirement for primary data collection through field mapping, even if only small-scale maps are required for planning purposes or the amount of background material is quite comprehensive.

Once the factual data have been collected and compiled then derivative maps or plans can be prepared

Table 2. *Types of applied geological map relevant to engineering*

Data points
- Location of exploratory holes and wells
- Distribution of geotechnical data test results
- Point rockhead information

Disturbed ground (human activity)
- Distribution (general)
- Distribution of mines and mine workings (all types, including surface and subsurface)
- Distribution of made-ground

Superficial geology
- Soil types, extent, lithology and thickness
- Drift thickness/rockhead contours
- Geotechnical properties of soils

Bedrock geology
- Rock types, extent, lithology, lithostratigraphy
- Structure contours
- Geological structure
- Geotechnical properties of rocks

Engineering geology
- Foundation conditions
- Hydrogeological conditions
- Ground conditions in relation to groundwater
- Nature and distribution of geohazards (subsidence, instability, flooding, earthquakes, etc.)
- Engineering geological zones (i.e. areas of homogenous engineering geological conditions)
- Aggregate and borrow material sources

Geomorphology
- Geomorphological landforms and process
- Drainage
- Areas of slope instability
- Flood frequency limits

Derived construction constraints maps
- Slope steepness
- Ground instability (e.g. subsidence, cambering, landslides, soft ground, etc.)
- Landslide hazard and risk maps
- Previous industrial usage (brownfield sites and contaminated land)

Derived resources maps
- Nature, extent and properties of mineral resources (superficial and bedrock)
- Groundwater resources
- Distribution of aggregates
- Sites of Special Scientific Interest (SSSIs)

Summary maps
- Development potential
- Summary of construction constraints
- Statutory protected land

Based on Smith & Ellison (1999).

that meet the specific requirements. Derivative maps are obtained by summarizing, simplifying or combining factual data and presenting them in a different form (see Edwards 2001). An indication of the wide range of maps found to be of value in engineering and planning situations is presented in Smith & Ellison (1999) based on the results of the UK Applied Geological Mapping Programme (Table 2).

Map legend

The map legend used on an engineering geological map should be developed to suit the purpose of the study. The general legend provided in the First Working Party Report on Maps and Plans (Anon. 1972), and presently being updated, remains the most useful basis for developing a task-specific legend. Dearman (1991) also contains a wide range of examples of map legends that have been used both in the UK and overseas. As with all legends it is critical that the symbol used or zone identified is fully explained. This might take the form of an expanded legend in the key or a short report, either accompanying the map or printed on the bottom of the map. As a general recommendation the map should be able to stand alone and be understood by all potential users without having to refer to a separate report.

Conclusions

Engineering geological mapping should be part of all site investigations and, in conjunction with other techniques of land surface evaluation, would form the basis for the development of the geological ground model, as proposed in Fookes (1997). The data from the mapping should be combined with ground investigation data and an updated ground model developed prior to construction. This model would then be further refined during construction as new soil and rock exposures become available.

References

ANON. 1972. The preparation of maps and plans in terms of engineering geology. *Quarterly Journal of Engineering Geology*, **5**, 293–381.
ANON. 1976. *Engineering Geological Maps: A Guide to their Preparation*. The UNESCO Press, Paris.
BARNES, J. W. 1997. *Basic Geological Mapping*. 3rd Edition. John Wiley & Sons, Chichester.
DEARMAN, W. R. 1991. *Engineering Geological Mapping*. Butterworth Heineman, Oxford.
DEARMAN, W. R. & FOOKES, P. G. 1974. Engineering geological mapping for civil engineering practice in the United Kingdom. *Quarterly Journal of Engineering Geology*, **7**, 223–256.
EDWARDS, R. J. G. 2001. Creation of functional ground models in an urban area. *This volume*.

FOOKES, P. G. 1969. Geotechnical mapping of soils and sedimentary rock for engineering purposes with examples of practice from the Mangla Dam project. *Géotechnique*, **19**, 52–74.

FOOKES, P. G. 1997. Geology for engineers: the geological model, prediction, and performance. *Quarterly Journal of Engineering Geology*, **30**, 293–424.

GRIFFITHS, J. S. 2001. Development of a ground model for the UK Channel Tunnel portal. *This volume.*

GRIFFITHS, J. S. & EDWARDS, R. J. G. 2001. The development of land surface evaluation for engineering practice. *This volume.*

GRIFFITHS, J. S. & MARSH, A. 1986. BS5930: the role of geomorphological and geological techniques in a preliminary site investigation. *In*: HAWKINS, A. B. (ed.) Site Investigation Practice: Assessing BS5930. Geological Society, London, Engineering Geology Special Publications, **2**, 261–267.

PETER, A. 1966. Essai de carte géotechnique. *Sols-Soils, Paris*, **16**, 13–28.

POPOV, I. V., KATS, R. S., KORIKOVSKAIA, A. K. & LAZAREVA, V. P. 1950. *Metodika sostavlenia inzhenernogeologicheskikhikart* (The Techniques of Compiling Engineering Geological Maps). Gosgeolizdat, Moscow.

SMITH, A. & ELLISON, R. A. 1999. Applied geological mapping for planning and development: a review of examples from England and Wales 1983 to 1996. *Quarterly Journal of Engineering Geology*, **32**, S1–S44

Landslide hazard mapping and risk assessment

G. J. Hearn[1] & J. S. Griffiths[2]

[1] Scott Wilson Kirkpatrick & Co Ltd, Basingstoke, Hampshire, UK
[2] Department of Geological Sciences, University of Plymouth, Devon, UK

Definitions

Before landslide hazard mapping and risk assessment are reviewed, it is important to define terms and concepts closely in order to avoid the confusion and misuse that has occurred in some previously reported case histories. The most widely accepted and basic definitions in landslide studies are those provided by Varnes (1984).

Natural Hazard: the probability of occurrence within a specified period of time and within a given area of a potentially damaging phenomenon.

Vulnerability: the degree of loss to a given element or set of elements resulting from the occurrence of a natural phenomenon of a given magnitude. It is expressed on a scale from 0 (no damage) to 1 (total loss).

Specific Risk: the expected degree of loss due to a particular natural phenomenon. It may be expressed by the product of *Hazard* and *Vulnerability*.

Elements at Risk: the population, properties, economic activities, including public services, etc., at risk in a given area.

Total Risk: the expected number of lives lost, persons injured, damage to property or disruption of economic activity due to a particular natural phenomenon. It is therefore the product of *Specific Risk* and *Elements at Risk*.

These definitions have been expanded by the International Union of Geological Sciences (IUGS) Working Party on Landslides through its committee on Risk Assessment (IUGS 1997) but the main elements are essentially the same.

Hazard, therefore, defines the *potential to cause damage*. With respect to landslides it is necessary to: identify the existence of a landslide or a potential slope failure; establish its size, depth, speed and travel distance; and estimate or calculate its frequency of movement or its probability of occurrence. **Risk** defines the *vulnerability* and *value* of the elements at risk; that is, establishing the *damage* or *loss potential* posed by the hazard should it occur. Risk is measured in terms of economic loss, hardship, loss of livelihood and threat to public safety. Thus, a phenomenon that poses a high hazard, such as a large, fast-moving landslide or debris flow, occurring on a remote mountainside may pose little risk, while a small fall of boulders in an urban area can have serious risk consequences for property and public safety.

Most maps published to date aim to identify and portray regions, areas or individual slopes that are likely to be more prone to failure than others. These **susceptibility maps** can vary significantly in scale, from the regional to the slope-specific. Landslide inventories and susceptibility maps produced for national or regional landslide registration and planning purposes are usually published at scales of 1:25 000 to 1:100 000, and occasionally smaller, while maps produced for local planning or engineering schemes are larger scale, usually between 1:5000 and 1:25 000. Maps may also range from a simple depiction of the known distribution of landslides, determined from aerial photographs and field mapping, to Geographical Information Systems (GIS)-based landslide susceptibility maps that analyse the conditioning and triggering factors that promote slope failure. These factors are usually evaluated from a combination of remotely sensed data, published information and field mapping and investigation. True **landslide hazard maps** that incorporate probability and the potential to cause damage, are less common. Further discussion and illustration of landslide susceptibility and hazard mapping techniques is contained in Hutchinson (1995) and Aleotti & Chowdhury (1999).

Landslide risk assessment attempts to combine the hazard and risk parameters into a framework allowing planning and project management decisions to be made. Quantification of these parameters (quantitative risk assessment or QRA) allows different options to be examined and compared, and the costs and potential benefits of hazard mitigation schemes or hazard avoidance strategies to be evaluated in terms of reduced risk, i.e. offset economic loss and improved public safety. A comprehensive review of this subject is contained in Cruden & Fell (1997).

Methodology

Landslide hazard mapping

Rather than describing a specific technique with a set procedure, **landslide hazard mapping** represents a range of

From: GRIFFITHS, J. S. (ed.) *Land Surface Evaluation for Engineering Practice*. Geological Society, London, Engineering Geology Special Publications, **18**, 43–52. 0267-9914/01/$15.00 © The Geological Society of London 2001.

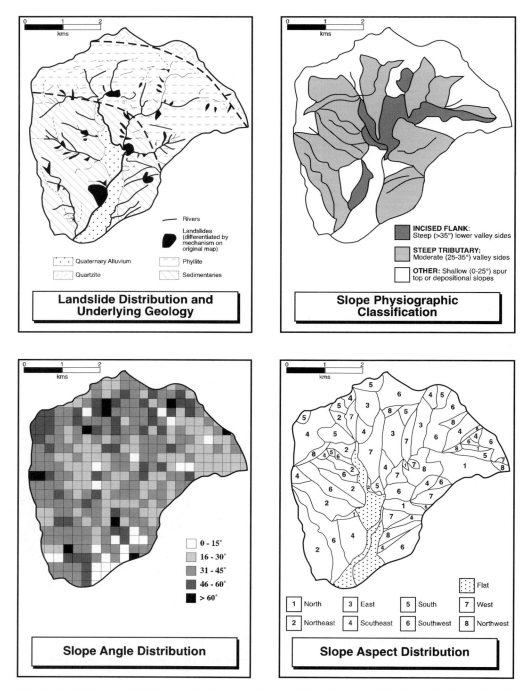

Fig. 1. Landslide susceptibility mapping for route alignment through an unstable river basin in east Nepal.

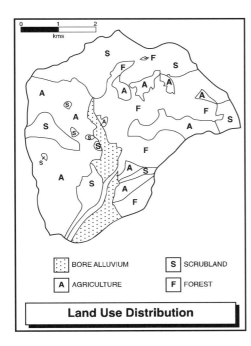

Land Use Distribution

BORE ALLUVIUM   **S** SCRUBLAND

A AGRICULTURE   **F** FOREST

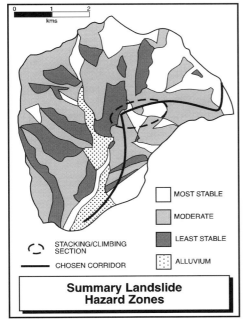

Summary Landslide Hazard Zones

MOST STABLE

MODERATE

LEAST STABLE

ALLUVIUM

STACKING/CLIMBING SECTION

CHOSEN CORRIDOR

FACTOR	CATEGORY	O / E	HAZARD RANK
ROCK TYPE	Sedimentaries	1.2	3
	Phyllite	0.8	2
	Quartzite	0.3	1
SLOPE ASPECT	North	0.7	1
	South	2.0	2.5
	East	2.2	3
	West	0.7	1
	Northwest	0.6	0
	Northeast	0.9	2
	Southeast	0.9	2
	Southwest	0.5	0
PHYSIO-GRAPHY	Incised flank	2.9	3
	Steep tributary	1.1	2
	Other	0.3	1
LAND USE	Scrub	1.0	N
	Agriculture	1.1	O
	Forest	0.9	T
SLOPE ANGLE	0 - 15°	0.7	S
	16 - 30°	1.0	I
	31 - 45°	1.3	G
	46 - 60°	0.8	N
	> 60°	0.4	I
CHANNEL PROXIMITY	Stream rank:		F
	First Order	0.9	I
	Second Order	1.1	C
	Third Order	1.5	A
	Fourth Order	0.9	N
	Fifth Order	1.0	T

Build up of hazard rank for the illustrated catchment. Hazard ranks for each factor category are summed for every terrain unit and assigned to one of three hazard classes. 3 is the most unstable condition.

N.B. The expected number of landslides (E) is calculated on the basis of the percentage study area coverage of each factor category multiplied by the total number of observed landslides (O) in the study. Thus, higher O/E ratios indicate a higher preponderance of instability than would be expected from a random distribution.

Fig. 1. (*continued*)

techniques that have been developed by different practitioners, each with different approaches, differing datasets with which to work and different specific mapping objectives. The unifying factor in virtually all attempts to portray landslide susceptibility or landslide hazard mapping is that one, or more, of the following assumptions is made.

1. The location of future slope failure or ground movement will be determined by the distribution of existing or past landslides, i.e. known landslide locations will continue to be a source of hazard.
2. Future landslides or ground movement will occur under similar ground conditions to those pertaining at the sites of existing or past landslides, i.e. the conditioning or controlling factors that gave rise to existing landslides can be ascertained and their distribution reliably mapped over wider areas. When examined collectively, these factors can provide a reasonable indication of the relative tendency for slopes to fail.
3. The distribution of existing and future landslides can be approximated by reference to conditioning factors alone, such as rock type or slope angle.

Mapping techniques that have relied on the first assumption are loosely classified as *direct mapping* methods in that they are based on a known distribution of existing landslides or slope failures. Methods that rely on inference (the third assumption) are classified as *indirect mapping*. These methods may rely on the knowledge or expectation that a given rock type is more prone to failure than another, while steeper slopes or the presence of an undercutting river will create a higher failure potential, other factors being equal. An approach based on the combination of the two methods (the second assumption) is most desirable, because a known landslide distribution will help to clarify and confirm the relative importance of controlling factors in landslide initiation. Furthermore, representation of the geographical variability in these controlling factors allows an interpretation of landslide susceptibility to be extrapolated into areas where, for whatever reason, failure has not yet occurred, or into areas where landslide event data are unavailable.

Landslide susceptibility mapping, therefore, is normally based on geographical correlation between landslide distribution and landslide controlling factors over large areas using existing data sources. Published geological and topographical maps are usually available, and together provide information on lithology, geological structure, slope steepness and aspect. Some may even show mapped landslide deposits. Aerial photographs and, to a lesser extent, satellite imagery can be useful in identifying landslides and slopes prone to failure. Furthermore, aerial photograph interpretation can assist with the identification of geographical patterns of landsliding, and the recognition and mapping of controlling factors. For example, colluvial deposits,

river undercutting, waterlogged ground and adverse rock structure orientations are also often identifiable on aerial photographs.

The manner in which the mapping procedure is set up has a significant effect on the landslide controlling factors that are selected, analysed and represented. Landslide susceptibility and hazard mapping is a terrain-based techniques of assessment and, as such, a study area should be subdivided according to slopes and terrain units for hazard analysis. The logical steps in any land surface evaluation involve the recognition of patterns in the geomorphological and geological landscapes, and in this respect landslide hazard mapping is no different. However, landslide hazard mapping based purely on statistical correlation between landslide occurrence and slope angle and rock type, for instance, has frequently been done on a grid square basis for ease of computer analysis. One of the earliest examples was that of Carrara *et al.* (1978) using land unit parcels of 200 m × 200 m. Notwithstanding the obvious value of this early work, the extent to which landslide controlling factors remain consistent over a 200 m square interval remains debatable.

Figure 1 shows a simple landslide susceptibility mapping exercise undertaken in 1984 for road alignment purposes in Nepal (Hearn 1987; TRL 1997). Slope angle was measured on a grid square basis from aerial photographs using a parallax bar. The grid cells were 230 m × 230 m and proved too coarse to allow slope angle to be significantly correlated against landslide distribution. A map that depicted the distribution of physiographical zones, identified and classified from aerial photographs according to slope morphology, proved far more appropriate (Fig. 1). One of the more recent examples is that of Greenbaum (1995) who used Landsat TM pixels (circa 30 m × 30 m) as the basic mapping unit. When landslide hazard mapping is contemplated at the regional and national scale, as in the cases described by Greenbaum, then the use of a grid square basis for data gathering and data analysis (*black box* method) becomes far less time consuming than a landform-based approach (*white box* method). It is preferable, with grid-based data, to employ GIS methods of landslide hazard data management, analysis and mapping output (Greenbaum 1995). Each data layer, such as topography or lithology, can be represented by an individual map. Algorithms are then used to undertake various spatial analyses on the data sets. Further discussion on this subject can be found in Soeters & van Westen (1996).

Hazard analysis and quantitative risk assessment (QRA)

Identifying which slopes are most prone to movement or failure is the first step in the assessment of landslide susceptibility hazard and risk. The next steps relate to the actual risk from an identified hazard and the need to address the questions: *How, When* and *With what effect*? In carrying out a quantified (landslide) risk assessment in this way, the following parameters need to be ascertained:

- the zone of influence of the hazard, i.e. its area of movement and its runout distance;
- the frequency of movement, e.g. the frequency with which failed masses travel over a given distance or range of distances, such as boulders impacting on buildings and public amenities located at varying distances from a rockfall source;
- the risk consequences attached to different event outcomes occurring, measured in economic terms, e.g. by reference to road maintenance costs, the costs of temporary road closures, the repair or reconstruction costs to damaged buildings and other infrastructure, and the cost of personal injury, fatality or loss of livelihood.

The following actions and input data are required before a landslide risk assessment, culminating in a QRA, can be fully achieved:

1. preparation of a landslide location and susceptibility map to identify potential landslide sources;
2. estimation of the volumes of potential landslide masses likely to be derived from these sources;
3. estimation of the likely areal influence and runout distance of these landslide masses;
4. assessment of frequency or return periods of different landslide and runout scenarios;
5. list of potential consequences and vulnerability of the elements at risk to these landslide and runout scenarios;
6. calculation of the economic loss and evaluation of the public safety implications associated with outcomes likely to take place over the lifetime of an existing or proposed development, for example 25 years in the case of a low-cost road, or 100 years in the case of a housing development;
7. if the risk levels calculated in action 6 are unacceptable, calculation of the cost of mitigation works and decision as to the most appropriate strategy for risk management. In the extreme, this may mean cancellation of the project.

In terms of mapping, landslide hazard assessment might succeed in delineating areas and frequencies of landslide runout, such as debris flows from a mountain slope onto a terrace below, or it may be able to portray the frequency of damage to roads as recorded in road maintenance records. Thus the first four actions identified above can be achieved if sufficient mapping data and records of past events exist (e.g. Lee 2000).

Event and decision tree analyses (Wu *et al.* 1996; Lee *et al.* 2000) can be used to define and assign probabilities to the various consequences (action 5). For example there may be a 100% probablility of a debris flow destroying a boundary fence, but only a 10% probability of the same landslide reaching a building, and only a 1% probability of fatality if the building collapses or a pedestrian is caught in the debris. Event probability is derived either from historical data (most rarely) or through computer simulation (most frequently). As with any probabilistic analytical procedure based on limited factual data, the retention of plausible argument in the final conclusion is paramount.

Undertaking actions 6 and 7 requires moving outside the normal area of expertise of the engineering geologist/ geomorphologist. Input from social scientists and economists is needed in order for the social and economic costs associated with different levels of risk (action 6) to be calculated. It is also necessary for the acceptable level of risk to be defined (action 7), and this is likely to be a function of government legislation, legal precedents and the general societal appreciation of natural and man-made risks. However, it is critical that engineering geologists and geomorphologists have an input in defining acceptable risks otherwise these may be set at impossibly high levels (Fell & Hartford 1997).

Limitations

The quality of the **landslide hazard mapping** output is dependent on that of the data used and the assumptions made in its derivation. At the site or project area scale (1:1000 to 1:25000), resources and practicalities will usually permit detailed mapping and geotechnical modelling to be carried out. As the size of the area to be covered by the mapping increases to the regional scale (1:25000 to 1:50000) the level of field-derived data input reduces, and the analysis becomes broader and more generalized, and may be limited to a simple overlay of geology with topography. Most landslide hazard mapping falls into the regional scale category, and is essentially a remote exercise, conditioned by the data available from existing sources. These data may often be incomplete, and may lack information on a slope parameter that is considered central to the initiation of landslides in that area. Without this information, landslide hazard mapping will be flawed from the beginning. Furthermore, certain controlling factors are often judged to be more important than others in causing slope failure, and weighting factors are assigned accordingly. The weighting system illustrated in Table 1 was applied to rock slope hazard mapping in Papua New Guinea on the basis of field observation (TRL 1997). The system was able to model reasonably well the distribution of recorded failures and it was therefore used for predictive purposes. The weighting system illustrated in Figure 1 for Nepal was also based on observed factor map correlations with landslide locations. The use of weighting factors without this observational support is not recommended.

Table 1. *Rock slope hazard rating for a site in Papua New Guinea*

	Hazard index
Factor 1: Slope angle irrespective of lithological dip	
Limestone slopes (based on graphical database)	
Angles <30°	0
Angles 31–44°	1
Angles 45–60°	2
Angles >60°	3
Weighting factor, 3	
Mudstone slopes (based on graphical database)	
Angles <25°	0
Angles 26–30°	1
Angles 31–35°	2
Angles >35°	3
Weighting factor 3	
Factor 2: Slope angle with respect to angle of unfavourable lithological dip	
Limestone and mudstone slopes	
Angle is at least 10° less than dip angle	0
Angle is at least 5° less than dip angle	1
Angle is10° greater than dip angle	2
Angle is >10° greater than dip angle	3
Weighting factor: 3 where dip direction is directly out of slope; **2** where dip direction is obliquely out of slope	
Factor 3: Adverse bedrock sequences (permeable overlying less permeable lithologies and/or strong overlying weak lithologies)	
None	0
Repetitive sequence of thin beds	1
Single sequence with thin reservoir or strong rock stratum	2
Single sequence with thick reservoir or strong rock stratum	3
Weighting factor, 2	
Factor 4: Faulted or sheared rock masses (only one of Factors 4 and 5 may be included in rating)	
None	0
Yes, but with favourable structural orientation	1
Yes, with unfavourable structural orientation	2
Weighting factor, 2	
Factor 5: Rock-mass dilation (only one of Factors 4 and 5 may be included in rating)	
Massive with closed joints	0
Massive with open joints or moderately fractured with closed joints	1
Highly fractured with open joints	2
Crushed and dilated (as in H = 1 or H = 2 for Factor 4)	3
Weighting factor, 2	
Factor 6: Drainage	
Well defined channelled runoff and dry slopes	0
No evidence of surface runoff, sinks or seepages	1
Sinkholes, internal drainage or ponding	2
Active springs and seepages	3
Weighting factor, 1 for limestone slopes and 2 for mudstone slopes	
Factor 7: Slope unloading by toe erosion or failure from below	
None	0
Localized and infrequent or due to ancient slope failure	1
Periodic or due to recent slope failure	2
Active	3
Weighting factor, 2	

* Scale: 0 = least hazard; 3 = greatest hazard.

A hazard map sets out to portray the nature and scale of a hazard. However, the vast majority of hazard maps only portray the relative tendency or **susceptibility** of slopes to fail. To indicate the level of hazard that might be generated they would have to incorporate frequency or probability and some indication of areal influence such as potential runout distances. These maps can, at best, portray relative hazard only; they do not provide an indication of factor of safety, nor do they usually give an indication of *how large*, or *how often*? Therefore, from a planner's or engineer's point of view most hazard maps will fall short of what is ultimately required for decision making, unless they are backed up by some form of risk assessment. If a hazard map is to progress from a document that displays relative hazard to one that provides an indication of absolute hazard, and from then to a document that portrays true risk, the quality and range of the necessary input data need to increase considerably.

Landslide hazard maps also rarely take into account the effects of triggering factors. Most landslides in mountainous regions are small shallow failures involving colluvium, soil or weathered rock. These landslides are usually triggered by intense rainstorms which, unless generated by demonstrably marked orographic effects, are just as likely to occur in one location as another over a long enough period of time. However, this is little help to a practitioner who has derived a landslide hazard map, only to find that the following year an intense localized rainstorm triggers numerous failures in an area where, either through lack of data or lack of precedent, a low or moderate hazard had been assigned. Earthquakes are other obvious triggers that tend to initiate or reactivate deeper failures. Whilst the location of active faults may be known, the process of converting seismicity into geographical landslide hazard is an uncertain one.

In hazard studies the timing of unknown failures will also usually be related to a triggering mechanism and is virtually impossible to evaluate except in terms of return periods, and even then usually with a considerable degree of uncertainty. Known failures and susceptible slopes, however, can be monitored to ensure there is pre-warning of accelerated movement, although the warning periods are often very short. For longer-term studies an indication of frequency might be gained from historical records, aerial photographs or geobotanical evidence, but this is very much dependent on the availability and relevance of these long-term records in the first place.

Finally, the same slope may pose a hazard from rock falls and debris slides or debris flows, but each may be conditioned by a different combination of controlling factors, and initiated by different trigger mechanisms and trigger levels. The hazard maps produced by Keinholz (1978) are amongst the few that have attempted to combine type of hazard with qualitative assessment of risk on the same map.

The main limitation with **landslide risk assessment** lies in the usual paucity of data with which to undertake a thorough analysis and reach an objective conclusion regarding the frequency of a landslide of a given size and a given runout distance, and the value of the impact consequence. It is important to ensure that engineering geological judgement prevails when numbers are being sought to satisfy the needs of the computer analysis. Putting values on potential for loss of land, loss of earnings, disruption to communications, damage to property, and personal injury and fatality is a difficult task. A community's perception and tolerance, or even acceptance, of disruption and risk to personal safety is hugely variable. This can lead to a disaster scenario being regarded as unacceptable in one country, while it is accepted as a fact of life in another.

Hong Kong is a useful example of what can be achieved. In Hong Kong, aerial photographs have been taken of the territory since 1942, and there are fairly comprehensive records of slope failures where these have taken place in areas of urban development and related infrastructure. QRA has therefore been possible in these circumstances (Malone 1998; Hardingham *et al.* 1998; Reeves *et al.* 1998). However, in areas where landslide events are less well documented and where less is known about landslide controlling factors and triggering mechanisms, the prospects for landslide QRA are far less.

Applications

Landslide hazard maps can be extremely valuable to regional planning and preliminary site selection, route alignment corridor identification, and in the provision of a framework for slope management and slope maintenance (TRL 1997). A register of large landslides – preferably GIS-based with an integrated attribute database describing their geology and geomorphology, probable cause, failure mechanisms, and extent and periodicity of movement – represents an extremely valuable source of information for planning purposes. The register of landslides recently compiled from aerial photographs in Hong Kong is a good example (Evans 1998; Griffiths *et al.* 1999). Taking the database a stage further into formal landslide susceptibility and hazard mapping also offers significant advantages if the factors controlling or triggering landslide location and behaviour can be adequately measured and subdivided geographically into meaningful landform elements. However, any inaccuracies in this process could blight areas of low hazard, and careful field verification is required in marginal zones.

In many applications it is not necessarily a problem that landslide hazard mapping and risk assessment fail, conceptually, to evaluate absolute hazard, multiple hazards and true risk. A combination of landslide susceptibility mapping and landslide runout analysis will often suffice when route corridors for roads and pipelines are being examined. This assumes that residual stability problems can be overcome or managed, the structure has to be built, and the objective is to find the most stable corridor in which to build it. High value elements of the construction, such as bridges and construction camps, need to be located at sites where landslide, flood and scour hazards are at a minimum, but these sites are usually selected on the basis of a detailed examination of candidate sites identified from engineering criteria. In the usual case where there is uncertainty over the magnitude, impact potential, timing or frequency of landslides or slope movements, it will always be preferable to use whatever mapping and analytical methods can be applied to better define landslide hazards on engineering geological grounds. If either the historical record, aerial photographs or field mapping indicate that large landslide events have taken place in the past under similar geomorphological and climatic conditions to those that pertain at present, then it has to be assumed that they will recur in the future. Their mitigation may be cost-ineffective in the case of low to moderate value investments but they will effectively become the 'design event' for such high value structures as expressways and housing developments.

In a recent study in Hong Kong (Hadley et al. 1998), geomorphological mapping was used to identify potential failure masses located on a granite mountainside above the proposed location for an expressway. A total of 43 potential landslide sources were identified and the volumes calculated from the mapping data. The runout geometry of a large debris flow that had occurred in 1990 was used to test a number of empirical runout formulae derived from other mountain regions and identify the one which was able to match, most closely, the 1990 Hong Kong case. All of the observed potential failure masses were routed down the main drainage lines below. The computed debris flow runouts reached the expressway embankments in only three of the 12 valleys, and in all cases there was sufficient embankment height to retain the anticipated volumes of debris. As a precautionary measure, checkdams were scheduled to be constructed in the steepest drainage lines upstream of the expressway alignment. This study serves to illustrate the case where the frequency or return period of a landslide event cannot be reliably assessed, and the risk assessment is therefore required to identify a design event that represents the realistic worst-case outcome.

Fig. 2. Slope treatment and road reconstruction across the Jogimara Landslide, Prithvi Highway, Nepal.

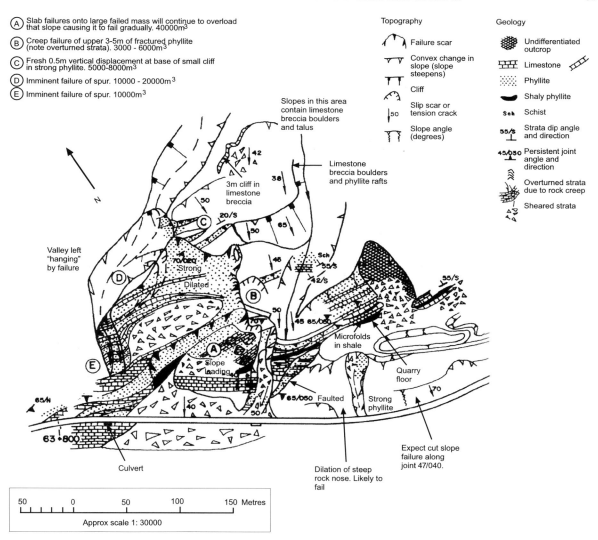

(A) Slab failures onto large failed mass will continue to overload that slope causing it to fail gradually. 40000m³

(B) Creep failure of upper 3-5m of fractured phyllite (note overturned strata). 3000 - 6000m³

(C) Fresh 0.5m vertical displacement at base of small cliff in strong phyllite. 5000-8000m³

(D) Imminent failure of spur. 10000 - 20000m³

(E) Imminent failure of spur. 10000m³

Topography

Failure scar

Convex change in slope (slope steepens)

Cliff

Slip scar or tension crack

Slope angle (degrees)

Geology

Undifferentiated outcrop

Limestone

Phyllite

Shaly phyllite

Schist

Strata dip angle and direction

Persistent joint angle and direction

Overturned strata due to rock creep

Sheared strata

Slopes in this area contain limestone breccia boulders and talus

Limestone breccia boulders and phyllite rafts

3m cliff in limestone breccia

Valley left "hanging" by failure

Strong Dilated

Microfolds in shale

Quarry floor

Slope loading

Faulted

Strong phyllite

Culvert

Dilation of steep rock nose. Likely to fail

Expect cut slope failure along joint 47/040.

50 0 50 100 150 Metres
Approx scale 1: 30000

Fig. 3. Engineering geological map of the Jogimara landslide showing anticipated failure volumes.

A study undertaken in Nepal illustrates the case where close to the worst-case outcome had already taken place. The main highway into Kathmandu was repeatedly blocked by rock falls from a particularly unstable rock slope during the monsoon seasons of the late 1980s and early 1990s (Fig. 2). Access would be disrupted for several days at a time and queues of traffic would build up for several kilometres on either side. Fatalities have been common. A combination of aerial photograph interpretation and engineering geological mapping assisted in identifying the main causes and mechanisms of failure. From the mapping, several zones of fractured, adversely jointed and overhanging rock were identified

and delineated. The volumes of these rock masses were estimated (Fig. 3) and time periods assigned during which their failure was considered most likely. These time periods ranged between 12 months and five to ten years. A number of stabilization and risk reduction options were considered, ranging from minor, essentially cosmetic, slope treatments and trap walls, through rock anchoring, bolting and shotcreting schemes to rock shelter and covered tunnel options for road and traffic protection. These options were costed and, given the relatively short time period in which the majority of remaining unstable rock masses were considered likely to fall, the recommendation was made to the client to adopt the option of minor slope treatment with a rock trap wall. Seven years later, the stability of the slope appears to have improved, and initial indications would therefore suggest that engineering geological mapping and judgement proved able to assist in a slope management decision where, through lack of data, quantitative risk assessment would have been difficult to apply.

Conclusions

Landslide hazard mapping now has an established 'track record' and, whilst most studies only evaluate relative rather than absolute hazard, their value is not questioned for many applications, particularly at the reconnaissance or planning level. Genuine risk assessment, however, is still in its infancy as far as application to landslide studies is concerned. The concept has been developed in far more detail in connection with the manufacturing and nuclear industries (Gerrard 2000) and was given strong impetus by the Royal Society Commission into risk analysis (The Royal Society 1992). With respect to landslides and indeed most natural hazards, with the exception of a few notable examples such as Hong Kong, the data requirements are presently far too onerous for full quantitative risk assessments to be normally undertaken. Clearly the need is for further development of data collection techniques to allow risk assessment of all natural hazards to be part of any terrain evaluation study.

References

ALEOTTI, P. & CHOWDHURY, R. 1999. Landslide hazard assessment: summary review and new perspectives. *Bulletin of Engineering Geology and the Environment*, **58**, 21–44.

CARRARA, A., CATALANO, E., SORRISO-VALVO, M., REALI, C. & OSSI, I. 1978. Digital terrain analysis for land evaluation. *Geologia Applicata e Idrogeologia,* **13**, 69–127.

CRUDEN, D. & FELL, R. (eds) 1997. *Landslide Risk Assessment.* Proceedings of the International Workshop on Landslide Risk Assessment, Honolulu, Hawaii. Balkema, Rotterdam.

EVANS, N. C. 1998. The natural terrain landslide study. *In*: LI, K. S., KRAY, J. N. & HO, K. K. S. (eds) *Slope Engineering in Hong Kong.* Balkema, Rotterdam, 137–144.

FELL, R. & HARTFORD, D. 1997. Landslide risk management. *In*: CRUDEN, D. & FELL, R. (eds) *Landslide Risk Assessment.* Proceedings of the International Workshop on Landslide Risk Assessment, Honolulu, Hawaii. Balkema, Rotterdam, 51–109.

GERRARD, S. 2000. Environmental risk management. *In*: O'RIORDAN, T. (ed.) *Environmental Science for Environmental Management.* Prentice Hall, Harlow, 435–468.

GREENBAUM, D. 1995. *Project summary report: Rapid methods of landslide hazard mapping.* Technical Report **WC/95/30**. British Geological Survey/DFID.

GRIFFITHS, J. S., HUNGR, O., HUTCHINSON, J. N., HARDINGHAM, A. D. & DITCHFIELD, C. 1999. *Scoping Study for a Global Quantitative Risk Assessment of Natural Terrain Landslides in Hong Kong.* Atkins China Ltd Report GEO 19/97 to Geotechnical Engineering Office, Hong Kong.

HADLEY, D., HEARN, G. J. & TAYLOR, G. R. 1997. Debris flow assessments for the Foothills Bypass, Hong Kong. *In*: LI, K. S., KRAY, J. N. & HO, K. K. S. (eds) *Slope Engineering in Hong Kong.* Balkema, Rotterdam, 153–162.

HARDINGHAM, A. D., DITCHFIELD, C. S., HO, K. K. S. & SMALLWOOD, A. R. H. 1998. Quantitative risk assessment – a case history from Hong Kong. *In*: LI, K. S., KRAY, J. N. & HO, K. K. S. (eds) *Slope Engineering in Hong Kong.* Balkema, Rotterdam, 145–151.

HEARN, G. J. 1987. *An evaluation of geomorphological contributions to mountain highway design with particular reference to the Lower Himalaya.* PhD Thesis, University of London.

HUTCHINSON, J. N. 1995. Landslide hazard assessment. *Proceedings of the Sixth International Symposium on Landslides*, Christchurch, New Zealand, **3**, 1805–1842.

IUGS 1997. Quantitative risk assessment for slopes and landslides – the state of the art. *In*: CRUDEN, D. & FELL, R. (eds) *Landslide Risk Assessment.* Proceedings of the International Workshop on Landslide Risk Assessment, Honolulu, Hawaii. Balkema, Rotterdam, 3–12.

KIENHOLZ, H. 1978. Maps of geomorphology and natural hazard of Grindelwald, Switzerland, scale 1:10000. *Arctic and Alpine Research*, **10**, 169–184.

LEE, E. M. 2000. The use of archive records in landslide risk assessment: historical landslide events on the Scarborough coast, UK. *In*: BROMHEAD, E., DIXON, N. & IBSEN, M.-L. (eds) *Landslides in Research, Theory and Practice.* Proceedings of the Eighth International Symposium on Landslides, Thomas Telford, London, 905–910.

LEE, E. M., BRUNSDEN, D. & SELLWOOD, M. 2000. Quantitative risk assessment of coastal landslide problems, Lyme Regis, UK. *In*: BROMHEAD, E., DIXON, N. & IBSEN, M.-L. (eds) *Landslides in Research, Theory and Practice.* Proceedings of the Eighth International Symposium on Landslides. Thomas Telford, London, 899–904.

MALONE, A. W. 1998. Risk management and slope safety in Hong Kong. *In*: LI, K. S., KRAY, J. N. & HO, K. K. S. (eds) *Slope Engineering in Hong Kong.* Balkema, Rotterdam, 3–17.

REEVES, A., CHAN, H. C. & LAM, K. C. 1998. Preliminary quantitative risk assessment of boulder falls in Hong Kong. *In*: LI, K. S., KRAY, J. N. & HO, K. K. S. (eds) *Slope Engineering in Hong Kong.* Balkema, Rotterdam, 185–191.

SOETERS, R. & VAN WESTEN, C. J. 1996. Slope instability recognition, analysis and zonation. *In*: TURNER, A. K. & SCHUSTER, R. L. (eds) *Landslides: Investigation and Mitigation.* Transportation Research Board Special Report **247**. National Academy Press, Washington DC, 129–177.

THE ROYAL SOCIETY 1992. *Risk: Analysis, Perception and Management.* Report of a Royal Society Study Group, The Royal Society, London.

TRL 1997. *Principles of Low Cost Road Engineering in Mountainous Regions.* TRL Overseas Road Note 16. Transport Research Laboratory, Crowthorne.

VARNES, D. J. 1984. *Landslide Hazard Zonation: A Review of Principles and Practice.* UNESCO, Paris.

WU, T. H., WILSON, H. T. & EINSTEIN, H. 1996. Landslide hazard and risk assessment. *In*: TURNER, A. K. & SCHUSTER, R. L. (eds) *Landslides: Investigation and Mitigation.* Transportation Research Board Special Report **247**. National Academy Press, Washington DC, 106–120.

Geomorphological mapping

E. M. Lee

Department of Marine Sciences and Coastal Management, University of Newcastle, Newcastle-upon-Tyne, UK

Introduction

There has been a long tradition of geomorphological mapping to support land use planning, especially in Poland (e.g. Klimazewski 1956, 1961; Galon 1962) and France (e.g. Tricart 1965; for a history of geomorphological mapping in Europe, see Verstappen 1983). The Geological Society Working Party Report on maps and plans (Anon. 1972) identified examples of geomorphological mapping that could be of use to engineers. However, the value of the technique was best highlighted by its application to road projects in unstable terrain in Nepal and South Wales during the early 1970s (e.g. Brunsden *et al.* 1975*a*, *b*; Doornkamp *et al.* 1979; Jones *et al.* 1983). The techniques have also been successfully applied to dryland problems (e.g. Brunsden *et al.* 1979; Bush *et al.* 1980, Doornkamp *et al.* 1980; Cooke *et al.* 1982, 1985; Jones *et al.* 1986), soil erosion (e.g. Morgan 1995) and river management (e.g. Doornkamp 1982; Richards *et al.* 1987). Despite these applications, BS5930, C*ode of Practice for Site Investigations* (British Standards Institution 1981), contains little reference to geomorphological maps (Griffiths & Marsh 1986) and the technique remains a marginal skill practised by a few experienced engineering geomorphologists. Examples of good mapping practice can be found within the references cited in this paper.

Methods

The style and format of a geomorphological map needs to reflect the nature of the environment and the problems that need to be addressed. The following are some of the more common types of map.

- Regional surveys of terrain conditions, either to provide a framework for land use planning (e.g. the 1 : 25 000 scale geomorphology map of the Torbay area (Doornkamp 1988) and the 1 : 50 000 scale resources survey maps of Bahrain (Doornkamp *et al.* 1980)) or as part of the baseline studies for environmental impact assessment (e.g. Lee 1999).
- General assessments of resources (e.g. the Bahrain Surface Materials Resources Survey (Doornkamp *et al.* 1980)) or geohazards (e.g. the investigation of ground problems and flood hazard in the Suez City area,

Egypt (Bush *et al.* 1980; Jones, 2001)); the delineation of gypsum-related subsidence problems in the Ripon area of the UK (Thompson *et al.* 1996); the assessment of landslide and erosion hazard at Ok Tedi copper mine, Papua New Guinea (Hearn 1995).
- Specific-purpose surveys to delineate and characterize particular landforms (e.g. the investigation of pre-existing landslide problems in and around the Channel Tunnel terminal area, Folkestone (Griffiths *et al.* 1995; Griffiths 2001); the assessment of coastal erosion hazards at Blackgang, Isle of Wight (Moore *et al.* 1998)).

There is no single approach to geomorphological mapping. The method chosen will generally reflect the nature of the problem to be solved, the resources available and, not least, the training and experience of the mappers. However, all maps should seek to subdivide the landscape into units with similar surface form, materials and process characteristics. At the smaller scales these units will, inevitably, be terrain models or land systems (see Phipps 2001; Fookes *et al.* 2001). Individual landforms or terrain units (e.g. escarpments, dune ridges, river channels, landslides) might be recorded on medium-scale maps. Landform elements or geomorphological units (e.g. individual landslide blocks, within-channel bars, gullys) might be recorded on large-scale maps.

The method of recording geomorphological information may change with map scale, but the basic approach to data collection should remain the same: mapping of surface form, description of materials and recording evidence of process. However, a key distinction must be made between desk-based approaches (i.e. aerial photograph interpretation and desk study) and field studies (i.e. morphological mapping, recording near-surface materials and evidence of surface processes).

Mapping surface form

At the smaller scales, topographic map contours can be used to subdivide the landscape into units, based on slope steepness (i.e. from the spacing of contour lines) and slope form (i.e. from the shape of the contours). However, at larger scales contours reveal little about landforms and their assemblages. The technique of *morphological mapping* is the most convenient and efficient

From: GRIFFITHS, J. S. (ed.) *Land Surface Evaluation for Engineering Practice*. Geological Society, London, Engineering Geology Special Publications, **18**, 53–56. 0267-9914/01/$15.00 © The Geological Society of London 2001.

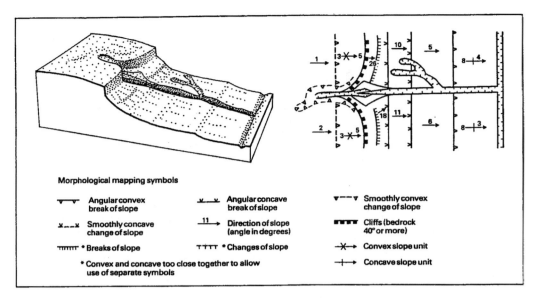

Fig. 1. Morphological mapping system (after Savigear 1965, from Cooke & Doornkamp 1990).

way of recording the surface morphology and allows later interpretation of form and process (Waters 1958; Savigear 1965). Breaks or changes of slope are identified from aerial photographs or in the field, and recorded using standard symbols (Fig. 1). The degree of generalization used in a morphological mapping survey will depend upon the scale of the base map used and the purpose of the exercise.

Morphological mapping in the field often involves two people and the use of a tape measure (25–50 m length) and compass. One person walks along a fixed traverse (oriented up and down the slope) with the tape until a break or change in slope is reached. The second person then records on the field map (i) the position (i.e. distance and compass bearing from the last point), (ii) the change in slope form (e.g. a convex change of slope) and (iii) the slope angle using a clinometer or Abney level, and 'fills-in' the map by joining this break or change in slope with the same feature recorded on previous traverses, adding to the map any features observed between the traverses. A useful rule is that all morphological lines must join up. The morphological map is built up through repeated traverses up and then down the slope (or vice versa). The distance between traverses will be dependent on the complexity of the landscape and the density of the vegetation: in dense undergrowth traverses need to be closer together than out in the open. A scale-rule, pencil, rubber (eraser) and sharpener are essential parts of the field kit, as ongoing modifications in the mapping are commonplace. It is pointless to continue mapping if you are unsure of your location: it will either lead to a poor map or large amounts of rubbing out.

Hand-held Global Positioning System equipment may be used, although it is of restricted value in densely vegetated terrain. In some instances it may be appropriate to record slope morphology 'by eye' or to use conventional survey equipment to accurately record the detail of surface features (e.g. using a series of surveyed markers).

Recording near-surface materials

The distribution of near-surface materials should be recorded at all exposures (e.g. landslide backscars, stream or wadi channels, cliffs, quarries or borrow pits), using standard methods of description (e.g. BS5930; Geotechnical Control Office 1988; Geotechnical Engineering Office 1996). The descriptions can be recorded in a notebook or on a pro forma developed or adapted for the mapping exercise. The second team member (i.e. the non-mapper) should be responsible for making the field notes or filling in the pro formas. It is essential to record the location of the observations. Photographs should be taken of the exposure and the photo number and subject recorded in the notebook or on the pro forma. Particular attention should be given to describing the nature of weathered materials and the texture of transported soils as these may give an indication of their origin and the nature and scale of the processes that occur within the landscape. For example, debris flow deposits consist of poorly sorted, large clasts embedded in a matrix of fine material. Boulders may be concentrated at the top of the deposit (i.e. reverse grading) because of the buoyant forces and dispersive pressures within a debris flow. By contrast, flash flood deposits tend to be cross-stratified and show fining-upward grading.

Recording evidence of surface processes

Evidence of surface process should be recorded (either active, recent or historic). Landsliding can be identified by the presence of characteristic landslide morphology or 'indicators of instability' (e.g. tension cracks, back-tilted trees, heave structures, etc.); floods or debris flows can be recorded by the upper limits of flow (e.g. from damaged tree trunks or strandlines) on the margins of floodplains; the net direction of sand transport can be identified from the build-up of material adjacent to groynes, nebkha (bush mounds in deserts) or windbreaks. The recording of surface drainage features, including areas of seepage, should form part of the mapping exercise. This is particularly important in areas of known or suspected instability.

Geomorphological map production

A geomorphological map is produced by an ongoing interpretation of the data collected in the field (or from satellite imagery or aerial photographs). A key stage in map creation is the identification of suitable mapping units to reflect the scale of mapping and the objectives of the study. Ideally, each unit should have consistent geomorphological characteristics, although internal variability of materials or rate of process may be a feature of the unit (e.g. in areas mantled by glacial tills). Three broad categories of geomorphological unit can be recognized:

1. units reflecting the control of the underlying geology (e.g. plateau surfaces, lithological benches, cliffs);
2. units reflecting the activity of surface processes (e.g. landslide, fluvial, aeolian features, etc.);
3. units reflecting modification of the landscape by man (e.g. areas of cut-and-fill, quarries, made-ground, etc.).

Geomorphological map unit boundaries should generally follow morphological boundaries, although some of the boundaries on the morphological map may be redundant. Depending on the map scale these units can be portrayed in blocks of contrasting colour or shading (as used on the 1:2500 scale geomorphological maps of the landslides at Ventnor, Isle of Wight (Lee & Moore 1991)) or as stylized symbols (see Demek 1972; Demek & Embleton 1978; Gardiner & Dackombe 1983; Cooke & Doornkamp 1990) or a combination of both.

Geomorphological maps should not be limited to a description of what has occurred; they should also be able to convey to the user what might occur, i.e. the potential for geohazards or sensitivity to change. For example, the combined body of evidence (site-specific and landform assemblages) can allow certain judgements to be made about the potential for landslide activity in the area. In terrain susceptible to debris flow activity, the mapping exercise should consider the way different landscape elements (e.g. hill slopes, valley floors and stream channels) interact to generate the potential for large flow events. In other circumstances, the recognition of clay-rich solifluction sheets indicates the potential for pre-existing shear surfaces that might be reactivated by inappropriate construction works.

Engineering geomorphology

The geomorphological map should provide a spatial framework for appreciating the processes and mechanics of landscape change over a range of timescales, particularly timescales relevant to engineers. The spatial framework provides 'context' for investigating problems. The old adage '*site and situation*' is clearly important in ensuring that potential geohazards are identified at an early stage in the investigation process (e.g. Fookes & Vaughan 1986; Fookes 1997). Hillside sites need to be seen in the context of the whole slope, river valleys as part of the whole catchment, coastal sites as part of a sediment transport cell (see Lee & Brunsden 2001).

Experience has shown (e.g. Cooke & Doornkamp 1990) that it is often necessary to produce some form of summary statement about the significance of the geomorphological conditions (landforms, near-surface materials and surface processes) to the engineering project for which it was undertaken. The summary information can be presented as:

*   a hazard map, e.g. showing the nature and distribution of landslide features within the area, susceptibility to landsliding, etc. (e.g. Hearn 1995);
*   a resource map, e.g. showing the distribution of aggregate resources (e.g. Cooke *et al.* 1982);
*   an extended map legend describing each map unit in terms that are relevant to the end user e.g. the landslide hazard or aggregate resource potential within each unit (e.g. the Ground Behaviour maps of the Ventnor area (Lee & Moore 1991)).

Geomorphological mapping, however, is only one of a number of complementary approaches which, when used in combination, can lead to the development of an effective geological model. The technique needs to be supported by, *inter alia*, subsurface investigation, monitoring, desk study, historical records analysis (e.g. sequences of aerial photos) and a review of the environmental controls, in order to fully establish the potential for geohazards or the availability of resources.

References

ANON. 1972. The preparation of maps and plans in terms of engineering geology. *Quarterly Journal of Engineering Geology*, **5**, 293–381.

BRITISH STANDARDS INSTITUTION. 1981. BS5930. *Code of Practice for Site Investigation.* British Standards Institution, London (updated in 1999).

BRUNSDEN, D., DOORNKAMP, J. C. FOOKES, P. G., JONES, D. K. C. & KELLY, J. M. N. 1975a. Large scale geomorphological mapping and highway engineering design. *Quarterly Journal of Engineering Geology*, **8**, 227–253.

BRUNSDEN, D., DOORNKAMP, J. C. HINCH, L. W. & JONES, D. K. C. 1975b. Geomorphological mapping and highway design. *Sixth Regional Conference for Africa on Soil Mechanics and Foundation Engineering*, 3–9.

BRUNSDEN, D., JONES, D. K. C. & DOORNKAMP, J. C. 1979. The Bahrain Surface Materials Resources Survey and its application to planning. *Geography Journal*, **145**, 1–35.

BUSH, P., COOKE, R. U., BRUNSDEN, D., DOORNKAMP, J. C. & JONES, D. K. C. 1980. Geology and geomorphology of the Suez city region, Egypt. *Journal of Arid Environments*, **3**, 265–281.

COOKE, R. U. & DOORNKAMP, J. C. 1990. *Geomorphology in Environmental Management.* Oxford University Press, Oxford.

COOKE, R. U., BRUNSDEN, D., DOORNKAMP, J. C. & JONES, D. K. C. 1982. *Urban Geomorphology in Drylands.* Oxford University Press, Oxford.

COOKE, R. U., BRUNSDEN, D., DOORNKAMP, J. C. & JONES, D. K. C. 1985. *Geomorphological dimensions of land development in deserts – with special reference to Saudi Arabia.* Nottingham Monographs in Applied Geography No. 4.

DEMEK, J. (ed.) 1972. *Manual of Detailed Geomorphological Mapping.* Academia, Prague.

DEMEK, J. & EMBLETON, C. (eds) 1978. *Guide to Medium-scale Geomorphological Mapping.* International Geographical Union, Brno.

DOORNKAMP, J. C. 1982. The physical basis for planning in the Third World. *Third World Planning Review*, **4**, 11–31.

DOORNKAMP, J. C. (ed.) 1988. *Planning and Development: applied earth science background, Torbay.* M1 Press, Nottingham

DOORNKAMP, J. C., BRUNSDEN, D., JONES, D. K. C., COOKE, R. U. & BUSH, P. R. 1979. Rapid geomorphological assessments for engineering. *Quarterly Journal of Engineering Geology*, **12**, 189–204.

DOORNKAMP, J. C., BRUNSDEN, D., JONES, D. K. C. & COOKE, R. U. 1980. *Geology, Geomorphology and Pedology of Bahrain.* GeoBooks, Norwich.

FOOKES, P. G. 1997. Geology for engineers: the geological model, prediction and performance. *Quarterly Journal of Engineering Geology*, **30**, 290–424.

FOOKES, P. G. & VAUGHAN, P. R. 1986. *A Handbook of Engineering Geomorphology.* Blackie, Glasgow.

FOOKES, P. G., LEE, E. M. & SWEENEY, M. 2001. Pipeline route selection and ground characterization, Algeria. *This volume.*

GALON, R. 1962. *Instruction to the detailed geomorphological map of the Polish Lowland.* Polish Academy of Science, Geography Institute of Geomorphology and Hydrography of the Polish Lowland at Torun.

GARDINER, V. & DACKOMBE, R. 1983. *Geomorphological Field Manual.* George Allen and Unwin, London.

GEOTECHNICAL CONTROL OFFICE. 1988. *Guide to Rock and Soil Descriptions.* Geoguide 3, Hong Kong Government.

GEOTECHNICAL ENGINEERING OFFICE. 1996. *Guide to Site Investigation.* Geoguide 2, Hong Kong Government.

GRIFFITHS, J. S. 2001. Development of a ground model for the UK Channel Tunnel portal. *This volume.*

GRIFFITHS, J. S. & MARSH, A. H. 1986. BS5930: the role of geomorphological and geological techniques in preliminary site investigation. *In*: Hawkins, A. B. (ed.) *Site Investigation Practice: Assessing BS5930.* Geological Society, London, Engineering Geology Special Publications, **2**, 261–267.

GRIFFITHS, J. S. BRUNSDEN, D., LEE, E. M. & JONES, D. K. C. 1995. Geomorphological investigation for the Channel Tunnel and Portal. *Geography Journal*, **161**, 257–284.

HEARN, G. J. 1995. Landslide and erosion hazard mapping at Ok Tedi copper mine, Papua New Guinea. *Quarterly Journal of Engineering Geology*, **28**, 47–60.

JONES, D. K. C. 2001. Ground conditions and hazards: Suez City development, Egypt. *This volume.*

JONES, D. K. C., BRUNSDEN, D. & GOUDIE, A. S. 1983. A preliminary geomorphological assessment of part of the Karakorum highway. *Quarterly Journal of Engineering Geology*, **16**, 331–355.

JONES, D. K. C., COOKE, R. U. & WARREN, A. 1986. Geomorphological investigation, for engineering purposes, of blowing sand and dust hazard. *Quarterly Journal of Engineering Geology*, **19**, 251–270.

KLIMASZEWSKI, M. 1956. The principles of geomorphological survey of Poland. *Przeglad Geograficzny*, **28** (Supp..), 32–40.

KLIMASZEWSKI, M. 1961. *The problems of the geomorphological and hydrographic map of the example of the Upper Silesian industrial district.* Polish Academy of Sciences Institute of Geography.

LEE, E. M. 1999. *In Amenas Gas, Algeria: baseline terrain evaluation report.* Confidential Report to Environmental Resources Management.

LEE, E. M. & BRUNSDEN, D. 2001. Sediment budget analysis for coastal management, Dorset. *This volume.*

LEE, E. M. & MOORE, R. 1991. *Coastal Landslip Potential: Ventnor, Isle of Wight.* Department of the Environment, London.

MOORE, R., CLARK, A. R. & LEE, E. M. 1998. Coastal cliff behaviour and management: Blackgang, Isle of Wight. *In*: MAUND, J. G. & EDDLESTON, M. (eds) *Geohazards and Engineering Geology.* Geological Society, London, Special Publications, **15**, 49–59.

MORGAN, R. P. C. 1995. *Soil Erosion and Conservation*, 2nd edn. Longman, Harlow.

PHIPPS, P. J. 2001. Terrain systems mapping. *This volume.*

RICHARDS, K. S., BRUNSDEN, D., JONES, D. K. C. & McCRAIG, M. 1987. Applied fluvial geomorphology: river engineering project appraisal in its geomorphological context. *In*: RICHARDS, K. S. (ed.) *River Channels: Environment and Process.* Blackwell, Oxford, 348–382.

SAVIGEAR, R. A. G. 1965. A technique of morphological mapping. *Annals of the Association of American Geographers*, **53**, 514–538.

THOMPSON, A., HINE, P. D. GREIG, J. R. & PEACH, D. W. 1996. *Assessment of subsidence arising from gypsum solution.* Symonds Travers Morgan Report to the Department of the Environment.

TRICART, J. 1965. *Principles et méthodes de la géomorphologie.* Masson, Paris.

VERSTAPPEN, H. TH. 1983. *Applied Geomorphology: Geomorphological Surveys for Environmental Development.* Elsevier, Amsterdam.

WATERS, R. S. 1958. Morphological mapping. *Geography*, **43**, 10–17.

Geographical information systems

C. P. Nathanail & A. Symonds

School of Chemical Environmental Engineering, Nottingham University, Nottingham, UK

Definition of Geographical Information Systems (GIS)

GIS software can be used to manipulate and display spatial information. Burrough & McDonnell (1998, p. 11) provide a number of definitions of GIS, based upon the concepts of the toolbox, database and organizational systems. Examples of each include:

- Toolbox definition: '. . . *a system for capturing, storing, checking, manipulating, analysing and displaying data which are spatially referenced to the earth . . .*'
- Database definition: '. . . *any manual or computer based set of procedures used to store and manipulate geographically referenced data . . .*'
- System definition: '*An organized collection of computer hardware, software, geographic data and personnel designed to efficiently capture, store, update, manipulate, analyse, and display all forms of geographically referenced information*'.
- Organization based definition: '. . . *a decision support system involving the integration of spatially referenced data in a problem solving environment . . .*'

In the context of terrain evaluation the 'organization based' definition is considered to be the most appropriate. This is because the intended output of terrain evaluation is to support a decision making process. More specifically, GIS can help the terrain evaluation process answer questions such as those in Table 1.

Table 1. *Spatial queries in terrain evaluation (Modified from Nathanail 1994)*

Where is the slope greater than x°?
What material is at x metres depth?
How far from the fault will the excavation be at x metres depth?
How many water tables are there and where are they?
What is the orientation of bedding across the site?
Where is the strength of clay less than x kPa?
Can bored piles be used here?
Can the sandstones in these boreholes be correlated?
How long will it take the water to flow from here to there?
Is more information needed; if so what information and where?

GIS data models

Present day GIS store information in both raster and vector formats. Raster data comprise 'pixels' or tiles that fill space, thus something is stored about everywhere. The raster data structure has its origins in remote sensing and digital image processing software. Vector data are stored as points, lines or polygons, thus something is stored only where a change occurs. The vector data structure has its origin in computer aided design (CAD) packages. Increasingly software allows users to store both types of data and to swap between one format and another for optimal data processing, display or reporting.

GIS uses and users

The core of a GIS is its database (Worboys 1995). Spatial data representing geographical phenomena are stored in this in terms of: their position with respect to a known co-ordinate system; their attributes that are unrelated to position; and their spatial interrelations (Burrough & McDonnell 1998). Through computer hardware components (e.g. monitor, keyboard, mouse, printer, plotter, scanner, digitizer (Burrough & McDonnell 1988; p. 12–14; Worboys 1997, p. 1–3; Martin 1996, p. 10–11)). GIS software is used to perform five generic functions on these data (Martin 1996; Burrough & McDonnell 1998):

(a) Data input and verification – the capture of spatial data and conversion to a digital form
(b) Data storage and database management – the way topology, attributes of geographical elements (e.g. points, lines, polygons) are structured and organized in the system
(c) Data output and presentation – the method of display and manner in which data are output to the user
(d) Data transformation – deals with the aspect of data correction and data analysis
(e) Interaction with the user – a range of menu driven, macro & programming language and 'hotkey' functions

From: GRIFFITHS, J. S. (ed.) *Land Surface Evaluation for Engineering Practice.* Geological Society, London, Engineering Geology Special Publications, **18**, 57–58. 0267-9914/01/$15.00 © The Geological Society of London 2001.

Table 2. *A summary of standard GIS functionality (after Symonds 1999)*

Function	Sub-sets of function
Capture	Digitizing/scanning Raster/vector conversion Co-ordinate/projection transformation Construct topology
Storage	Relational/object-oriented databases Data integration
Retrieval	Data browsing Windowing Query generation
Manipulation	Map generalization Map abstraction Reclassification Scale change Linear and rubber sheeting
Analysis	Statistical summary Buffer generation Polygon overlay/dissolve Measurement – distance areas, volumes Least-cost routes Network analysis Spatial/environmental modelling Surface modelling and analysis Map algebra
Display	Visualization Multiple map views Multiple feature displayed Tabular, chart, graphic output

From: G3 Solutions 1999.

Worboys (1995) suggests that there are a number of analytical requirements for a GIS: resources inventory; network analysis; terrain analysis; layer based analysis; location analysis; spatial-temporal information. Martin (1996, p. 59–60) adds reclassification and neighbourhood characterization. He also suggests that sequences of these functions are known as cartographic modelling. Within most GIS are a large range of subsets of each of these forms of analysis (Table 2).

This functionality leads GIS to have many potential uses and users. For example, uses include: environmental modelling (Strachan & Stewart 1996); forestry, land registry, utilities, transport and engineering applications (Tomlinson 1987; Martin 1996); business strategy development (Hendriks 1998); business applications (90% of business users data spatially related, Grimshaw 1994).

As the price of computer systems and GIS fell in the 1980s and 1990s, the range of users increased considerably: emergency services; regulators; insurers; financial institutions; estate agents; surveyors; environmental groups; and, property developers (DOE 1994). As if to confirm its wide appeal, Burrough & Frank (1995) have described GIS as a *'generic toolbox ...'*, with the ability to solve many different problems for different user types.

The development in the resolution of global positioning systems and the drop in their price means there is no excuse for not collecting accurate locations of where observations were made or samples taken. Without such location information, GIS can be of no use.

At the time of writing desk top GIS have power and capability that would have required workstations only a couple of years ago. The key to any GIS implementation remains an adequately populated and maintained database, well trained staff and operating procedures that explicitly take the presence of GIS into account.

References

BURROUGH, P. A. & FRANK, A. U. 1995, Concepts and paradigms in spatial information: are current geographical information systems truly generic? *International Journal of Geographical Information Science*, **9**, 101–115.

BURROUGH, P. A. & MCDONNELL, R. A. 1998. *Principles of GIS*. Oxford University Press, Oxford.

DEPARTMENT OF THE ENVIRONMENT. 1994. *Contaminated Land Research Report: Information systems and land contamination*, CLR Report No. 5, DOE, London.

G3 SOLUTIONS 1999. *The Countryside Information System Year 2000 Project*, available at: http://g3solutions.freeserve.co.uk

GRIMSHAW, D. J. 1994. *Bringing Geographic Information Systems into Business*, Longmans.

HENDRIKS, P. H. J. 1998. Information strategies for GIS. *International Journal of GIS*, **12**, 621–639.

MARTIN, D. 1996. *Geographic Information Systems: socioeconomic applications*, 2nd edn. Routledge, London.

NATHANAIL, C. P. 1994. *Systematic modelling and analysis of digital data for slope and foundation engineering*. PhD Thesis, London University.

STRACHAN, A. J. & STUART, N. 1996. UK Developments in Environmental GIS. *International Journal of GIS*, **10**, 17–20.

SYMONDS, A. 1999. *Developing a contaminated land inspection strategy for local authorities using a Geographical Information System*. MSc Thesis, University of Nottingham.

TOMLINSON, R. F. 1987. Current and potential uses of geographical information systems: The North American experience. *International Journal of GIS*, **1**, 203–218.

WORBOYS, M. F. 1997. *GIS: A Computing Perspective*. Taylor & Francis, London.

Terrain systems mapping

P. J. Phipps

Mott MacDonald, Croydon, Surrey, UK

Description

The term 'terrain systems mapping' originated in the 1930s and early 1940s when the requirement was identified to classify large areas of terrain for the purposes of locating potential agricultural and economic resources, and identifying suitable sites for development in mainly undeveloped rural areas (Mitchell 1973). The view is adopted in this paper that 'terrain' and 'land' are synonymous, which follows both Christian & Stewart (1968) and Townshend (1981).

In many respects the methodology behind terrain systems mapping has not altered since one of its first significant applications came to prominence in Australia, with the publication of a series of reports summarizing terrain systems mapping of the Australian Territories (Christian & Stewart 1952). Areas up to hundreds of square kilometres were delineated in which characteristic assemblages of topography, soils and vegetation could be identified. This association is described in Cooke & Doornkamp (1990) after Stewart & Perry (1953). These authors established that the topography and soils are dependent on the nature of the underlying rocks (geology), the erosional and depositional processes that have produced the present topography (geomorphology) and the climate under which these processes have operated. Thus the land system is a scientific classification of country or landscape based on topography, soils and vegetation correlated with geology, geomorphology and climate.

The basic good practice for developing a suite of terrain systems maps has been succinctly covered in the 1982 Engineering Group Working Party Report. The three main landform units of terrain system, terrain facet and terrain element are defined and described. Terrain systems maps are hierarchical to an extent which depends on the scale adopted, mapping requirements and brief (Fig. 1). The recognizable pattern forming the terrain systems have normally been mapped at a scale of 1:250 000 to 1:1 000 000. However, recent experience from studies for the Channel Tunnel Rail Link has shown that the application of mapping at 1:50 000 scale for civil engineering projects is most suitable when integrating detailed satellite imagery and re-evaluating geological mapping (Waller & Phipps 1996).

A terrain system covering a large area can be subdivided into a number of defining terrain facets such as a flat-topped interfluve or major valley which assist in describing the terrain system. It is the terrain facet that provides the basic landform unit. There will undoubtedly be variations in the attributes of a terrain facet, although the surface materials, underlying geology, water table, slope inclinations and slope stability will be relatively uniform within certain definable boundaries. Terrain facets have typically been mapped at 1:10 000 to 1:60 000 scale although Waller & Phipps (1996) have again shown their applicability at larger scales of 1:2500 to 1:5000.

The terrain facets may themselves be further broken down to terrain elements which are features that could reasonably be identified and mapped directly in the field. A terrain facet that describes a major valley could comprise terrain elements such as a stream channel, flood plain, ox-bow lake etc. (Fig. 1).

The most significant advances in terrain systems mapping since the last working party report have not affected the general approach, but the techniques for the establishment of hierarchical models have improved. The widespread availability of satellite imagery and high-powered desk-top computing facilities have enabled initial delineation of terrain systems boundaries to be carried out accurately and rapidly. Further detailed divisions at the terrain facet scale by aerial photograph interpretation or photogrammetric techniques can then be augmented by other Earth science information layers which could include vegetation, ground investigation information and hydrogeological readings. All the data can be managed and presented through computerized geographical information systems. Geostatistical techniques are available which can be applied to the information to provide, for example, levels of confidence for potential soil thicknesses in a specific terrain facet.

However, there is still a fundamental requirement in understanding the various geological, geomorphological and climatological processes that combine to sculpt the terrain of any area. Site reconnaissance and/or ground truthing are required to provide the basis for delineating terrain systems and terrain facets, and in validating the models and maps that are developed. Without the fieldwork element the terrain systems approach is not

From: GRIFFITHS, J. S. (ed.) *Land Surface Evaluation for Engineering Practice*. Geological Society, London, Engineering Geology Special Publications, **18**, 59–61. 0267-9914/01/$15.00 © The Geological Society of London 2001.

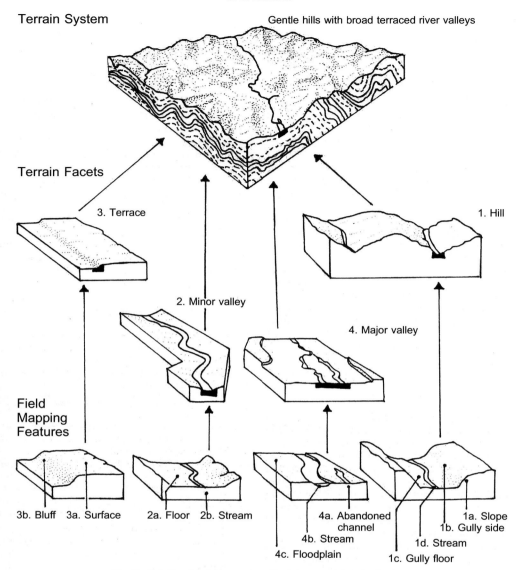

Fig. 1. Relationship between terrain systems and terrain facets.

rigorous and can lead to misinformation on a large scale that can be very costly to end users.

Limitations

The most significant limitation in carrying out a terrain systems mapping programme is related to the quality and nature of the base information. System and facet boundaries can only be delineated where distinct and visible on remotely sensed imagery whether satellite mounted or from aerial photography. The resolution, wavelength sensitivity, processing and enhancement will all affect the specific nature of images that are to be used for interpretation. Different images will highlight different characteristics of the ground surface which will affect visual distinctions. Remote sensing surveys commissioned specifically for a mapping study will need to be specified with much care. Furthermore, the availability of existing topographical and geological maps, ground investigation data and reports, and any other potential supporting information will assist in developing more representative mapping schemes and ground models at an early stage. These can then be augmented by targeted ground investigations, detailed field mapping or remote surveys to improve the information level in less well understood terrains.

The detail and appropriateness of the terrain system maps produced are also dependent on the skill and experience of the person undertaking initial interpretations, and information review and synthesis. An interpreter experienced in a specific terrain or environment should be able to provide higher information levels and summarize the characteristics of facets in a more efficient and cost-effective way than someone who is not familiar with that environment.

Finally, the scales adopted for terrain systems mapping are such that even though large areas can be mapped rapidly and information summarized on a terrain facet level, detailed information at a site-specific scale is often not available. To design a particular engineering structure or earth structure, civil engineers and geotechnical engineers require information on ground conditions and geotechnical parameters specific to the location of the structure. The variability inherent within a terrain facet will normally be too great to allow all but the most provisional of designs to be developed.

Applications

By far the most appropriate and effective use of terrain systems mapping is for feasibility studies related to long linear civil engineering structures that include railways, roads, water supply pipelines, canals and to a lesser extent tunnels. The maps and supporting information produced can not only assist in identifying potential alignments, but also provide information on engineering geological and geotechnical issues such as topography, bedrock geology, soils, groundwater, geomorphological and geological hazards, aggregate resources, excavatability, fill potential, foundation and slope stability. This suite of information topics is dependent on engineering requirements and the specific environment. Identification of facets with high saline potential for roads would be important in desert systems and the depth of residual soil development and weathering profiles in tropical environments for excavatability of pipeline trenches. Published examples of the terrain systems mapping approach for long linear structures are numerous, but some recent applications would include low-cost roads in the Himalayas (Fookes 1997), a high-speed rail link in southeast England (Waller & Phipps 1996), and road design in the Libyan desert (Hunt 1979).

Within the above context of civil engineering applications, terrain system maps can be developed over a wide areal extent for assisting in identifying locations for potential dam sites, quarries, airfields and large structures.

Terrain systems mapping has also been adopted to assist in a wide variety of applications that are not directly related to engineering issues. A summary of these has been provided by Mitchell (1973). Such usages include soil science, agriculture and forestry, meteorology, microclimatology and hydrology, and resource analysis for landscape and recreational planning. Terrain systems maps have also been used for military purposes in determining suitability of off-road mobility for different vehicles, identifying sites of relatively easy excavatability and for construction materials.

References

CHRISTIAN, C. S. & STEWART, G. A. 1952. *Summary of General Report on Survey of Katherine-Darwin Region 1946*. Land Research Series, 1, CSIRO, Australia.

CHRISTIAN, C. S. & STEWART, G. A. 1968. Methodology of Integrated Surveys. Proceedings of Conference on Aerial Surveys and Integrated Studies, Toulouse, Unesco, 233–280.

COOKE, R. U. & DOORNKAMP, J. C. 1990. *Geomorphology in Environmental Management*. Clarendon Press, Oxford.

ENGINEERING GROUP WORKING PARTY 1982. Land surface evaluation for engineering practice. *Quarterly Journal of Engineering Geology*, **15**, 265–316.

FOOKES, P. G. 1997. Geology for engineers: the geological model, prediction and performance. The First Glossop Lecture. *Quarterly Journal of Engineering Geology*, **30**, 293–424.

HUNT, T 1979. Geotechnical aspects of road design in Libya. *Ground Engineering*, October, 15–19.

MITCHELL, C. W. 1973. *Terrain Evaluation*. Longman, London

STEWART, G. A. & PERRY, R. A. 1953. *Survey of Townsville–Bowen Region (1950)*. Land Research Series 2, CSIRO, Australia.

TOWNSHEND, J. R. G. (ed.) 1981. *Terrain Analysis and Remote Sensing*. George Allen & Unwin.

WALLER, A. M. & PHIPPS, P. J. 1996. Terrain systems mapping and geomorphological studies for the Channel Tunnel Rail Link. *In*: CRAIG, C. (ed.) *Advances in Site Investigation Practice*. Proceedings of the International Conference, London, 30–31 March 1995. Thomas Telford, London, 25–38.

Section 3

Case Studies in Land Surface Evaluation

Surface and groundwater resources survey in Jordan

R. J. Allison

Department of Geography, University of Durham, Durham, UK

Purpose of survey

Water is one of the most valuable physical resources in the arid zone. Much effort is made by engineers to maximize availability and minimize wastage. In the Hashemite Kingdom of Jordan water scarcity has been exacerbated in recent years by rapidly rising demand. In the early 1990s total national water consumption in Jordan approached $730 \times 10^6 \, \mathrm{m}^3$. It is estimated that demand will rise to $1200 \times 10^6 \, \mathrm{m}^3$ by the year 2000.

In the northeast Badia of Jordan there are two major sources of water (Al-Homoud *et al.* 1995). Due to the topographic effect of the Jebel Druz, precipitation totals exceed $500 \, \mathrm{mm} \, \mathrm{a}^{-1}$ in the north, declining to $<50 \, \mathrm{mm} \, \mathrm{a}^{-1}$ in the south. During winter months, runoff can be considerable but no reasonable data exist on parameters such as wadi discharge, infiltration rates and drainage basin contributing areas under storms of a given magnitude. Groundwater is found in three aquifers. Numerous government-operated and private wells have recently

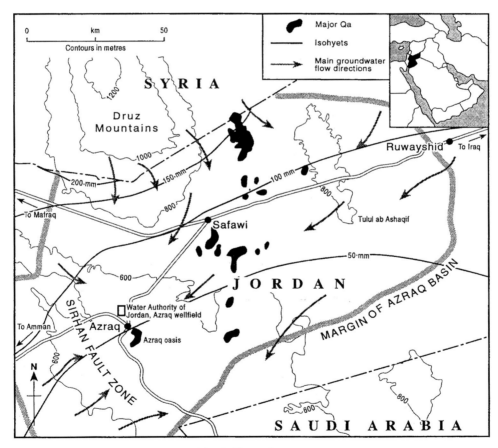

Fig. 1. The study area.

From: GRIFFITHS, J. S. (ed.) *Land Surface Evaluation for Engineering Practice*. Geological Society, London, Engineering Geology Special Publications, **18**, 65–71. 0267-9914/01/$15.00 © The Geological Society of London 2001.

been drilled to exploit groundwater. There is little information on the recharge:extraction balance, changing spatial patterns of water availability and temporal changes in water quality, despite trends which hint at a depleting resource.

The purpose of the research was to provide data for enhancing efficient use of a scarce water resource. Supporting objectives for the surface water study included determining water availability, establishing patterns of runoff generation, quantifying sediment mobilization and transport rates and locating potential water harvesting sites. Supporting objectives for the groundwater study included determining the rate of aquifer drawdown and recovery as a consequence of pumping, quantifying any permanent fall in water table height and aquifer storage, identifying the causes of aquifer pollution and establishing patterns of groundwater quality deterioration. Interaction between the surface and groundwater components of the project is significant. Understanding the overall hydrological regime requires synthesis of both parts of the study and these must be based on the development of a suitable geological and geomorphological ground model.

The site

The northeast Badia encompasses 11 200 km^2 of land (Fig. 1), around 14% of the total land area of Jordan. The spatial limits of the study broadly coincide with the margin of late Tertiary and early Quaternary basalt lava flows (Burdon 1959; Bender 1974, 1975), which spread from local eruptive centres (Fig. 2). The resulting basalt plateau is between 50 km and 170 km wide from east to west and 180 km from north to south. The basalts are alkaline–olivine in character (Ibrahim 1992). Absolute age determination using K-Ar techniques has identified a number of basalt lava flows, with ages ranging from 13.7 Ma to less than 0.5 Ma for the exposed rocks. Some of the unexposed flows which constitute part of the groundwater system date at 23 Ma.

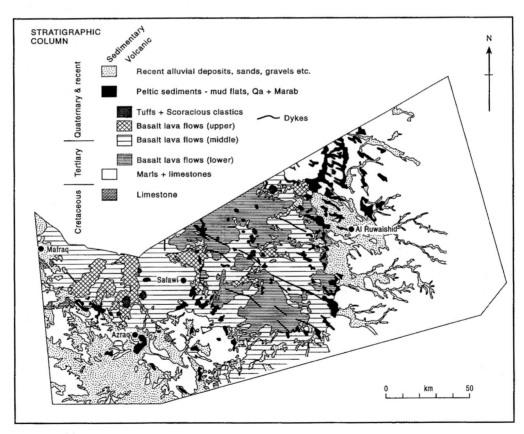

Fig. 2. Geological map and stratigraphic column of the northeast Badia (source: Bender 1974).

Topography rises from a low of 400 m in the south to 1200 m around the town of Mafraq. There is an accompanying increase in rainfall, which is seasonal and often concentrated in high magnitude/low frequency runoff generating storms. The region is dominated by low, gently undulating hills and elevation differences of 25 m to 30 m between high and low points across the landscape. Gradients are seldom steep and there are few sudden breaks of slope. Topographic highs provide local watersheds, particularly where they occur in lines along dykes or major fault systems. Regional structure is relatively simple. Three dominant fault systems run east to west, northwest to southeast and east-northeast to west-southwest. Structural lineaments have some control on groundwater flowpaths.

The age of the basalt flows determines the degree of drainage network development, most readily observed by wadi connectivity. The oldest lava flows have a gentle, rounded topography, well developed wadi systems and a fine colluvial rill network. The most recently emplaced basalts have a more rugged topography and poorly developed drainage network. As the basalts have weathered since emplacement, boulder fields have evolved, with clasts covering the ground surface and overlying fine-grained, light orange sediments. The size of clasts and the degree of protection which they afford to the ground surface varies. The Abed weathers to large, rounded boulders, with exposed bare ground. The Bishriyya produces a fine reg of basalt chips, which leaves very little of the underlying ground surface exposed. The ground surface boulder cover affects surface runoff and sediment mobility.

Different stratigraphic units act as either aquifers or aquitards, depending on their physical properties (Fig. 3). The lower aquifer occurs at depths of 1.3 km to 3.4 km,

is poorly understood, hydrothermal and heavily sulphurous. The middle aquifer dips beneath the study area at depths of 400 m to 700 m. Its transmissivity is low at 35 m² d to 450 m² d and water quality is variable. The upper aquifer is the most important groundwater resource. It consists of recent sediments, Tertiary basalts and limestones and is separated from the poorer quality middle and lower aquifers by a thick marl aquitard. The elevation of the upper aquifer is little more than 50 m in the north to around 350 m near Safawi. Flow is radial towards the Azraq basin, with the saturated thickness decreasing from >300 m in the north to <50 m in the south. Pump tests show transmissivity to be highly variable but exceeding 3000 m² d in some places.

Available information

A desk study at the start of the project revealed a considerable amount of data, much as unpublished government reports. The northeast Badia has been topographically mapped at a scale of 1:50 000. Remote sensing imagery is available including swaths of Landsat TM and SPOT digital data and air photographs. The Geology Directorate of the Natural Resources Authority have mapped stratigraphy and structure at a scale of 1:250 000. Subsurface geophysical surveys have been completed, revealing detail about the structure and the groundwater reserve.

Weather records confirm that much of the year is dominated by low precipitation and high potential evapotranspiration (Korzon 1974). Precipitation seldom exceeds 500 mm a⁻¹ and potential evapotranspiration can approach 2000 mm a⁻¹, a consequence of mean

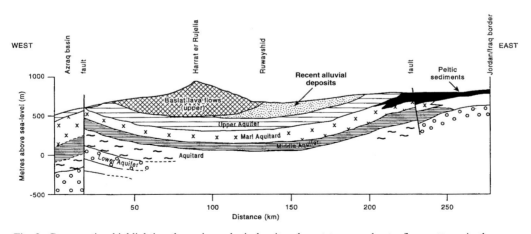

Fig. 3. Cross-section highlighting the main geological units relevant to groundwater flow patterns in the northeast Badia.

annual maximum temperatures of 34°C to 37°C. The Water Authority of Jordan monitors the three groundwater systems but principally the Upper aquifer. Huntings Technical Survey has undertaken soil mapping and analysis as part of the National Soils and Land Use Project.

A number of the information sources comprise incomplete data sets. There is no guarantee that a complete groundwater sampling programme will be undertaken during site visits, for example, and time-series data include breaks in the record. Variations in the hydrological conditions both at and beneath the ground surface have been so rapid in recent time that there is doubt as to whether the resolution of available information is adequate for establishing patterns of change.

Techniques used

Surface hydrology runoff pathways were established by digitizing 1:50 000 topographic maps. Contour patterns and drainage networks were recorded. The Geographical Information System package ARC-Info was used to examine the hydrological regime. An automatic weather station and data logger were installed towards the middle of the study area to measure precipitation, wind speed and direction, wet and dry bulb temperature, solar radiation and the net radiation balance. The data were used to establish the hydrological flux and balance between times of net water surplus and deficit. Field infiltration tests were undertaken using a constant head, double-ring infiltrometer. Tests were conducted on a variety of surfaces to partition the landscape into zones of high, medium and low infiltration.

Surveys in wadis and across their bounding slopes used an electronic distance measurer. Wadi gradients were determined for channel reaches at different points along drainage networks. Clasts were sampled and measured along their a- b- and c-axis for palaeohydrological reconstruction. Lines of painted stones of differing size were installed along wadi cross-sections. Sites were revisited to see whether flow competence during wet seasons reached magnitudes sufficient to move clasts, thereby permitting discharge calculations. Sequences of slope profiles were surveyed between topographic highs and lows to establish gradient, length and runoff potential.

Groundwater studies included tests undertaken at the well-head and samples collected for laboratory analysis. All samples were collected from purged boreholes. A stable well-head electrical conductivity was used to confirm adequate pre-sample pumping. An electronic water level probe was used to establish the height of the water table and pump tests were undertaken to quantify the rate of draw-down and recovery. Total dissolved

Table 1. *Details of wells which are characteristic of Upper and Middle aquifer water extraction sites in the northeast Badia*

Well no.	Location Eastings Northings	Aquifer type and lithology	Pump rate $(m^3 s^{-1} h^{-1})$	Conductivity mmmho/cm	Temperature (°C)	Dissolved oxygen (%)	pH	Total cations $(meq\, l^{-1})$	Total anions $(meq\, l^{-1})$
1	North of region Druze foothills 316570 184875	Basalt	58	300	29.7	88	8.42	2.79	2.81
2	Lower Druze foothills 322300 174500	Basalt and rijam limestone	45	610	34	87	8.14	5.14	5
3	Azraq wellfield north 321508 152443	Basalt and rijam limestone	180	610	26	100	8.28	5.51	5.05
4	Azraq town 321425 143800	Basalt	10	920	23.2	90	7.15	18.21	18.05
5	Central Azraq basin 320750 138900	Rijam limestone	n/a	1400	24.5	59	7.47	12.05	11.73
6	Eastern margin of Azraq basin 338200 140550	Basalt	60	1770	22.2	25	7.12	17.47	17.77
7	South-west margin of Azraq basin 316420 130033	Azraq formation	30	2140	23.5	89	7.05	23.89	23.05

oxygen was monitored during pump tests using a meter with an attached thermometer to correct for temperature. Electrical conductivity and pH were recorded. Samples of water were collected for well-head titration tests to determine alkalinity. Water (50 ml) was titrated against 0.16 N or 1.6 N H_2SO_4 using phenolphthalein indicator if CO_3^{2-} anions were present and screened methyl orange to determine the concentration of HCO_3^- anions. Bottles of water were returned to the laboratory for detailed geochemical analysis. Two 30 ml, filtered samples were collected at each site. One sample was acidified with three drops of 50% HCl. A 50 ml sample was collected for isotope analysis and at six sites a sample was collected for ^{14}C dating.

Conceptual model

The northeast Badia is an environment where water is limited and demands on the resource are increasing. Developing technologies permit surface water harvesting and groundwater extraction in increasingly large volumes, which cannot be sustained. Alterations to one component of the hydrological regime will affect other parts of the system.

The surface water regime is characterized by long periods of the year with a negative hydrological balance. Precipitation-induced runoff results in wadi flow, with flood hydrographs passing rapidly through the drainage network. Groundwater extraction is increasing, particularly from the upper aquifer, where the water table is close to the ground surface. Access to groundwater is determined by the thickness of permeable strata and the regional dip. The present rate of water use cannot be maintained. Both quality and volume of the upper aquifer are declining and action needs to be taken to halt over-exploitation.

What the survey established

Somewhere between 85% and 92% of precipitation is lost to evaporation, 5% to 11% is lost to infiltration and 2% to 4% generates runoff. Figures are highly seasonal. There are periods of the year during the spring, summer and autumn when the hydrological flux is in deficit. Much of the groundwater used for crop irrigation evaporates rather than infiltrating into the upper soil layers for plant uptake. During the wet season rainstorm events frequently generate overland flow. In places where the basalt boulder ground cover leaves exposed areas, sediment movement can be significant.

Ground surface hydrology is affected by the degree of drainage network development on basalt lava flows of different ages. Pans act as collection zones for water and sediment. On the youngest basalts, surface water flow is highly localized. Water runs into small, closed depressions or pans known as qa. The qa are usually no more than 100 m^2 or so in size and seldom linked. On older basalts there are qa which are much larger in size, some exceeding 20 km^2. The pans are fed from extensive areas and are often supplied by wadis as well as overland flow. Large qa have high surface evaporation rates and are usually saline, limiting the value of water once it has entered the depression. A third type of pan, known locally as marab, evolves where sediment is deposited along the course of wadis if channelled systems open out across wide areas. Marab gradients encourage the movement of water from their up-stream to their downstream end. Most are fed by an extensive up-basin network of wadis, resulting in significant water inundation during wet seasons.

Wadis drain radially into the Azraq basin. Wadi gradients range from 5% to 10% on the footslopes of the Druze and 3% to 5% in the south of the region. During the study wadi flow was limited and many of the cross-section painted boulder lines did not move. Palaeohydrological analysis of wadi bed deposits confirmed that significant discharge levels do occur and there is potential to use surface water more efficiently. Infiltration rates are generally high in wadi beds but decrease rapidly on qa, marab and towards interfluves. Where a combination of high infiltration rates and appropriate structural controls exist, there is the potential to supplement groundwater recharge.

Groundwater studies confirm that the total available resource and its quality are declining. Recharge to the upper aquifer is mainly through direct infiltration in the northern part of the basin and through wadi beds. Contemporary recharge estimates vary between 22 and 36 $Mm^3 a^{-1}$. Water quality in the upper aquifer is generally good but salinity levels are increasing. They range from a few hundred to 4000 ppm. At some well-heads the total dissolved solids is approaching a point where the water becomes marginal for human consumption, principally because of $NaHCO_3$ and NaCl levels (Fig. 4). Extraction is largely uncontrolled from many of the region's 600 boreholes, 75% of which are unlicensed. High evaporation rates where farmers are over-watering, in combination with decreasing resource quality, is leading to ground surface salinization. Poor well-head completion often leads to contamination. Lead concentrations in the south, for example, are the result of fuel spills and ingress down boreholes.

The general conclusion is that the hydrological regime in the northeast Badia is a function of surface water dynamics on the one hand and the groundwater regime on the other. Both are linked. Modifications to one component of either system have broad consequences. Expedient and integrated use of surface and groundwater from the upper aquifer has the potential to support sustainable development. The regional water authority

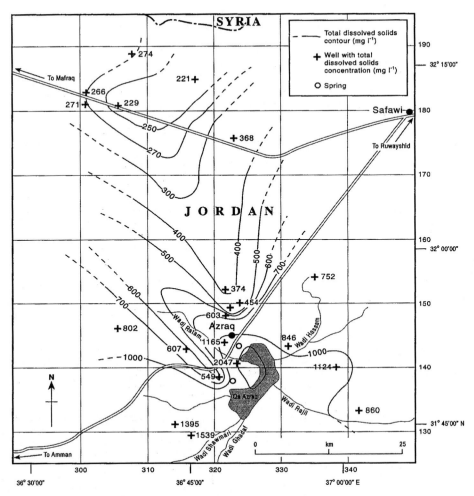

Fig. 4. Contour map of total dissolved solids of the upper aquifer complex.

plan is for the borehole network to the upper aquifer to be extended and abstraction raised by a further 3.9 × 10⁶ m³. Such a development, with no further integration of the surface and groundwater regimes, is likely to generate a gradually declining water resource in a hydrologically sensitive area.

Similar applications

Aspects of the wadi surveys would not be possible in perennial channels and issues such as throughflow in the unsaturated zone and the effects of vegetation would have to be considered in many environments. Examples

of similar studies in dryland environments include work in Syria and countries of the Arabian Peninsula. Surface water surveys have been completed in Bahrain as part of its earth surface and materials resources survey (Brunsden et al. 1979; Doornkamp et al. 1980). Wadi flow regimes have been examined for their potential to maximize available water in the Sultanate of Oman (Doyel et al. 1984; Maizels & Anderson 1988). Work has been completed on runoff dynamics and wadi discharge during high magnitude/low frequency events in the Negev desert (Schick 1977; Yair 1983, 1992) and parts of semi-arid United States of America (Patton & Baker 1977; Abrahams et al. 1992).

The groundwater programme is transferable to most aquifers. Sampling frameworks have to be designed relative to individual aquifer systems. The time between sampling and analysis is an important issue where stored water is likely to deteriorate. Examples of similar studies in other dryland environments include work in the Sultanate of Oman (Jones *et al.* 1988), Saudi Arabia (Bakiewicz *et al.* 1982; Lloyd & Pim 1990) and Syria (Khouri 1982). There are also studies in other parts of Jordan, particularly in the south of the country (Burdon 1982; Charalambous 1990).

Acknowledgements. The work presented here was undertaken as part of the Jordan Badia Research and Development Programme, jointly sponsored by the Royal Geographical Society (with the Institute of British Geographers), London, and the Higher Council for Science and Technology, Amman. Support was generously provided by the University of Durham Research Initiatives Fund. David Drury and Beatrice Gibbs completed parts of the groundwater survey and their input to the Programme is gratefully acknowledged.

References

ABRAHAMS, A., PARSONS, A. & HIRSCH, P. 1992. Field and laboratory studies of resistance to inter-rill overland flow on semi-arid hillslopes, southern Arizona. *In*: PARSONS, A. & ABRAHAMS, A. (eds) *Overland Flow – Hydraulics and Erosion Mechanics*. UCL Press, London, 1–23.

AL-HOMOUD, A. S., ALLISON, R. J., SUNNA, B. F. & WHITE, K. 1995. Geology, geomorphology, hydrology, groundwater and physical resources of the desertified Badia environment in Jordan. *GeoJournal*, **37.1**, 51–67.

BAKIEWICZ, W., MILNE, D. M. & NOORI, M. 1982. Hydrogeology of the Umm Er Radhuma aquifer, Saudi Arabia. *Quarterly Journal of Engineering Geology*, **15**, 105–126.

BENDER, F. 1974. *Geology of Jordan*. Gerbruder Borntraeger, Berlin.

BENDER, F. 1975. *Geology of the Arabian Peninsula: Jordan*. United States Geological Survey Professional Paper **560-I**.

BRUNSDEN, D., JONES, D. & DOORNKAMP, J. 1979. Bahrain Surface Materials Resources Survey and its application to planning. *The Geographical Journal*, **145**, 1–35.

BURDON, D. J. 1959. *Handbook of the Geology of Jordan*. Government of the Hashemite Kingdom of Jordan, Amman.

BURDON, D. 1982. Hydrogeological considerations in the Middle East. *Quarterly Journal of Engineering Geology*, **15**, 71–82.

CHARALAMBOUS, A. N. 1990. *Hydrogeology of the Disi Sandstone Aquifer*. UNDP/DTCD Project JOR/97/003, Hashemite Kingdom of Jordan Water Authority.

DOORNKAMP, J. C., BRUNSDEN, D., JONES, D. K. C. & COOKE, R. U. 1980. *Geology, Geomorphology and Pedology of Bahrain*. Geo Books, Norwich.

DOYEL, W. W., AUBEL, J. W., DAVISON, W. D., GRAF, C. G., JONES, J. R. & KENNEDY, K. G. 1984. *The Hydrology of the Sultanate of Oman*. Public Authority for Water Resources Report No. 83.1, Muscat.

IBRAHIM, K. M. 1992. *The geological framework for the Harrat Ash-Shaam basaltic super-group and its volcanotectonic evolution*. Natural Resources Authority, Amman.

JONES, J. R., WEIER, H. & CONSIDINE, P. R. 1988. Geology and hydrogeology of the pre-dune sand deposits of the Wahiba Sands, Sultanate of Oman. *Journal of Oman Studies*, **3**, 61–73.

KHOURI, J. 1982. Hydrogeology of the Syrian steppe and adjoining areas. *Quarterly Journal of Engineering Geology*, **15**, 135–154.

KORZON, V. I. 1974. *Atlas of World Water Balances*. Hydromet, Moscow.

LLOYD, J. & PIM, R. H. 1990. The hydrogeology and groundwater resources development of the Cambro-Ordovician sandstone aquifer in Saudi Arabia and Jordan. *Journal of Hydrology*, **121**, 1–20.

MAIZELS, J. & ANDERSON, E. W. 1988. Surface water in the Sharqiyah: flash floods February/March 1986. *Journal of Oman Studies*, **3**, 217–230.

PATTON, P. C. & BAKER, V. R. 1977. Geomorphic response of central Texas stream channels to catastrophic rainfall and runoff. *In*: DOEHRING, D. O. (ed.) *Geomorphology in Arid Regions*. London, Allen & Unwin, 189–217.

SCHICK, A. P. 1977. A tentative sediment budget for an extremely arid watershed in the southern Negev. *In*: DOEHRING, D. O. (ed.) *Arid Geomorphology*. John Wiley & Sons Ltd, New York, 139–163.

YAIR, A. 1983. Hillslope hydrology, water harvesting and areal distribution of some ancient agricultural systems, northern Negev. *Oecologia*, **47**, 83–88.

YAIR, A. 1992. The control of headwater area on channel runoff in a small arid watershed. *In*: PARSONS, A. J. & ABRAHAMS, A. D. (eds) *Overland Flow – Hydraulics and Erosion Mechanics*. UCL Press, London, 53–68.

Mapping for high pressure gas pipelines in South Wales

G. P. Birch

Consultant Engineering Geologist, Sevenoaks, Kent, UK

Objectives

In order to mitigate against environmental impact, to satisfy the design criteria and to ensure the long-term integrity of the construction, the Transmission Department of Wales Gas sought geotechnical advice on the routing and design of a major gas transmission system. This system was being built to reinforce the gas supplies to the industrial valleys of South Wales. The opportunity was taken to apply geomorphological mapping at an early stage in route planning so as to avoid abortive design work on alignments which might subsequently prove unsuitable or too costly to engineer (Fig. 1). Pipeline engineers welcomed the approach, which commenced by obtaining an understanding of the client's objectives and industry design guidance. Working as part of the project team, engineering geomorphologists provided guidance from initial routing studies through detailed design to construction, which was completed without contractual conflict arising out of unforeseen ground conditions.

The Project

Gas supplies to the industrial valleys of South Wales developed in piecemeal fashion outwards from the individual coking plants set up by collieries in the valley floors. The introduction of natural gas from the North Sea in the 1970s provided the opportunity to invest in a new transmission system emanating from the new high-pressure grid on the north edge of the Coal Field basin.

The 'backbone' of the project is a 30 km north–south, high-pressure, welded steel transmission/storage pipeline (oversized up to 1200 mm diameter to permit storage by pressure-packing) linking the North Sea feeder main at Dowlais, near Methyr Tydfil, with the Cardiff feeder main at Nantgarw, near Caerphilly (Fig. 2). Congestion within the intensely developed valley floor dictated a mountain route along the valley divide (Fig. 3), with smaller diameter cross-valley spurs to connect into the existing low-pressure distribution system by way of decompression stations.

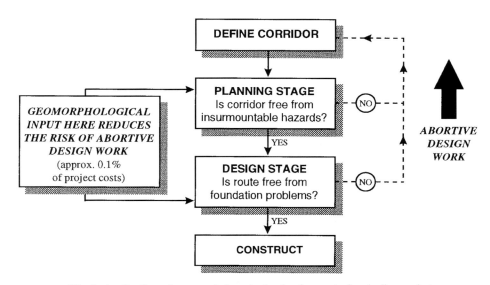

Fig. 1. Application of geomorphology to the development of a pipeline project.

From: GRIFFITHS, J. S. (ed.) *Land Surface Evaluation for Engineering Practice*. Geological Society, London, Engineering Geology Special Publications, **18**, 73–82. 0267-9914/01/$15.00 © The Geological Society of London 2001.

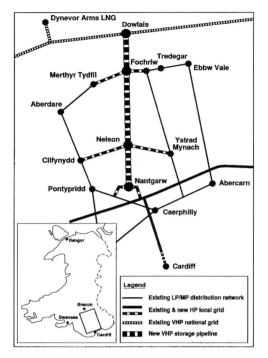

Fig. 2. Project layout.

Initial observations at the desk study stage revealed that the routes for both N–S and E–W pipelines were required to negotiate a wide range of topographic, geomorphological, geological and man-made constraints as illustrated in the 'generic' block diagram (Fig. 4).

Of particular significance to the project design was the large diameter relative to pipeline wall thickness and the high pressure to be contained up to 70 bar. The sensitivity of the 'thin wall' pipeline to tensile strain and, more especially, compressive strain required particular attention with regard to route selection and potential for ground movement.

Techniques

At the *planning stage*, the emphasis was on the use of existing information, such as geological maps and stereo air photographs, to assess proposed alignments (Fig. 5). This 'remote sensing' was particularly valuable where the developer was not in a position to obtain access for field surveys. The main objectives were first, recognition of surface form (morphology) and its relationship to geology, and second, the identification of the origin of the features (morphogenesis) and their stage of development in relation to time (Brunsden *et al.* 1975). The latter, in particular, provided the key to predicting hazardous ground conditions or processes which had the

Fig. 3. The 1200 mm diameter pipeline between Dowlais and Nelson was constructed in widely varying ground conditions ranging between soft swampland, where a flotation jacket was required, to some of the UK's strongest rock strata, where blasting was required.

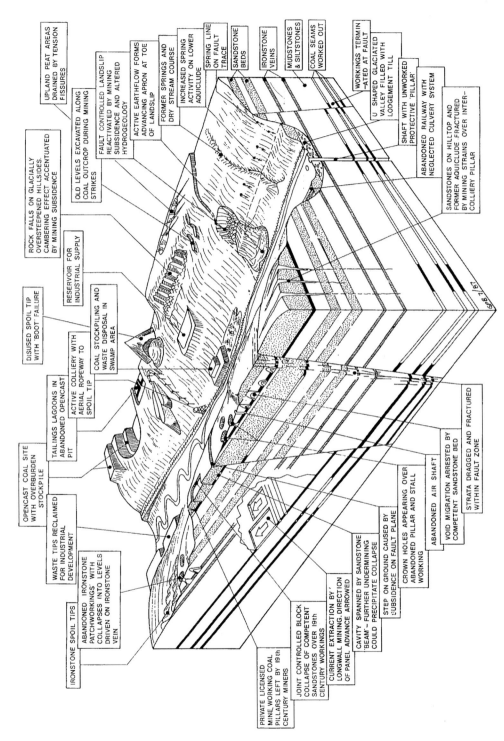

Fig. 4. Block diagram of a typical South Wales coalfield valley showing natural and man-made hazards and their significance to engineering projects.

GEOTECHNICAL INPUT - PIPELINE PLANNING STAGE

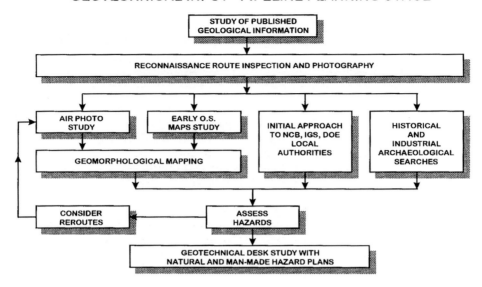

GEOTECHNICAL INPUT - PIPELINE DESIGN STAGE

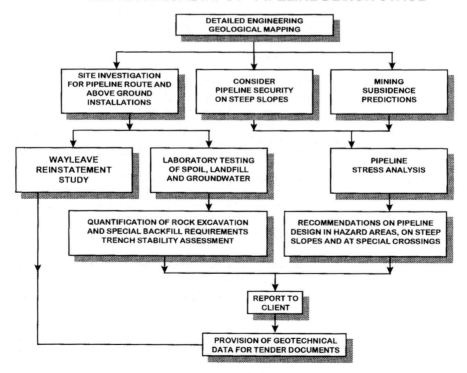

Fig. 5. Geotechnical input to pipeline planning and design.

FEATURE	KEYS TO RECOGNITION ON AERIAL PHOTOGRAPHS
NATURAL HAZARDS	
landslides	• arcuate backscar above concave slump • back-tilted strata • hummocky ground surface • leaning trees and patchy vegetation • disturbed field boundaries • changes of slope characteristics
erosion	• steep, freshly exposed river banks • light, striations and stripped vegetation • debris slides on lower valleyside • deeply incised ravines • ground loss downslope of springlines
cavernous ground	• subsidence hollow • disappearing and re-emerging drainage
marshland or peat	• lush vegetation and tonal patterns • differential reflectance between stereo images
geological fault	• surface lineation with fresh fissure or step if activated by undermining • displaced hedges or walls
MAN-MADE HAZARDS	
old shallow mining	• irregular hollows or "crown holes" • regular joint-controlled collapsed areas • tension fissures along slope crest
deep mining	• fissure swarms & fault steps
backfilled quarries	• unnatural and patchy vegetation patterns • absence of mature trees • removed fencelines
erosion	• drainage and erosion anomalies

Fig. 6. Selected natural and man-made hazards.

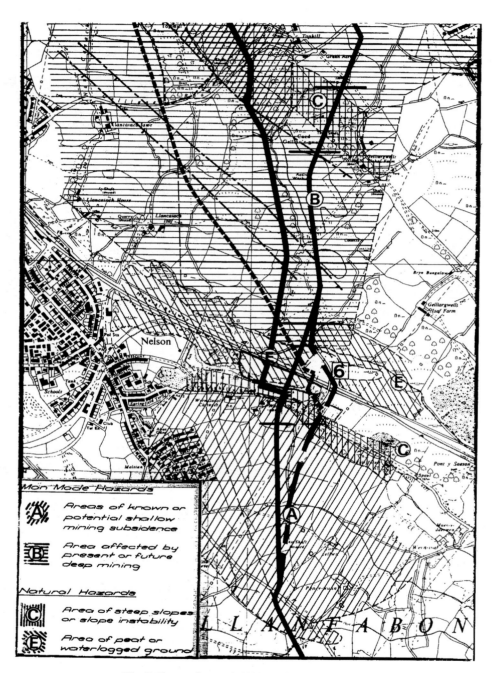

Fig. 7. Extract from a 1:10 000 scale hazard map.

Fig. 8. Subsidence collapses or 'crown holes', are common in areas of abandoned shallow mining. Their identification by geomorphological mapping provides an essential component of the hazard map.

potential to be reactivated during the construction period or subsequently during the life of the project.

The geomorphological and man-made features were recognized by a methodical study of the various elements of the landscape within, and in the vicinity of, the proposed route corridors. Air photos, viewed stereoscopically, provided a wealth of information on slope morphology, drainage patterns, former land use and variations in soil type and wetness, enabling the development of a model of the pattern of landforms and development processes (Fig. 6).

The features observed were plotted onto a series of 1:10 000 scale Ordnance Survey route maps which were then developed into a set of hazard maps by highlighting those features which had the potential to impact on the design, construction or operation of the project (Fig. 7).

Annotated in this way, the geomorphological map provided the pipeline engineer with a visually clear, low-cost and rapid evaluation of the route characteristics and likely ground conditions. Whilst this information could be provided by air-photo interpretation alone, it was desirable to carry out 'ground proofing' by field reconnaissance, when areas of potential concern could be assessed before consideration was given to subsurface investigations, rerouting or special design.

Once a preferred route had been established, the geomorphological map provided the basis for the design of subsurface investigations necessary for pipeline design. These comprised trial pits, boreholes and geophysical techniques. By distinguishing on the map between those areas displaying essentially uniform or benign geomorphological characteristics and those areas where difficult or hazardous ground conditions were indicated, it was possible to optimize the layout of trial pits to maximize their value and save on their overall numbers.

The 1:10 000 scale hazard mapping defined a number of areas where realignments were recommended to ensure the long-term integrity of the pipelines, for example, to avoid areas prone to collapse of shallow nineteenth century mine workings (Fig. 8).

At the *design stage*, detailed mapping at 1:2500 scale was used to update the geomorphological information which, combined with the subsurface information, facilitated an engineering geological evaluation of the proposed route (Fig. 5). By considering each element of the project and its engineering requirements, the air-photo, surface-photo and mapping information was used to assess the likely impact of the pipeline construction on the terrain (Hadley 1991). This provided an indication of any special design requirements, such as pre-contract drainage in wet areas, or heavy-wall pipe thickness in

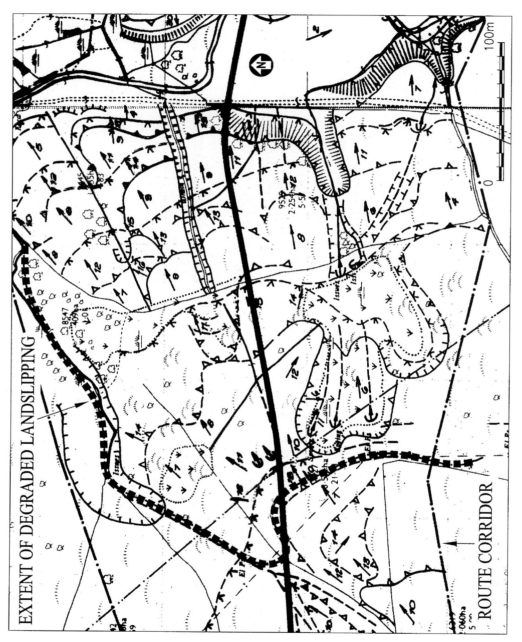

Fig. 9. Extract from 1:2500 geomorphology map.

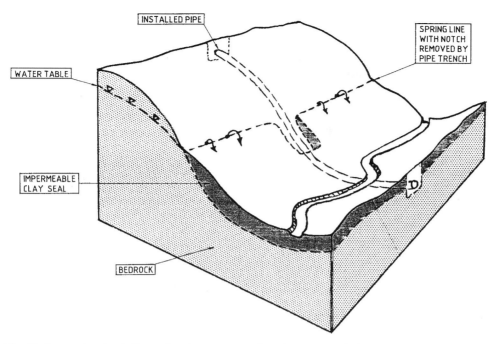

Fig. 10. Perspective sketch illustrating the permanent alteration to groundwater movement resulting from pipeline construction.

areas liable to ground movement or collapse. Not all hazard areas were avoidable and detailed geomorphological mapping enabled pipelines to be routed through extensive zones mapped by the BGS as undifferentiated landslipping (Fig. 9).

At the *construction stage*, rapid response was provided to resolve geomorphological or geotechnical problems relating to ancillary works or design alterations. Visits during construction provided the opportunity to validate the predictions made at the planning and design stages.

Comparisons

The following two examples are included to illustrate how geomorphological mapping might have averted costly post-construction remedials. The first, in Kent, is where a pipeline was constructed along a line running obliquely up a hillside with a history of instability. As a result, the pipe was subjected to high stresses from the ground attempting to carry it off downslope. The pipe was exposed for stress relief and monitoring and subsequently abandoned.

The second illustration is in Wales where the pipeline trench penetrated an impermeable clay seal of natural boulder clay lining the floor of the valley (Fig. 10).

Whilst posing no particular problems to construction during summer months, it became evident once the groundwater table rose again that the pipeline trench had permanently altered the local groundwater regime, focusing spring water into the backfilled trench. As a consequence, the trench backfill was blown out leaving the pipeline suspended and extensive underdraining work was necessary to reinstate the pipeline safely.

Other applications

The techniques for using engineering geomorphology in developing hazard assessments for linear development have been applied to a wide range of engineering projects, including, *inter alia*, the following:

- canal routing in dynamic environments (Peru) – determining viability prior to site survey (Birch 1989);
- roadway repairs on unstable land (UK) – prioritization of maintenance resources;
- railway construction audit (Spain) – identification and evaluation of problem zones and their causes (see Birch 2001);
- shallow tunnelling for urban metros (Greece) – determination of pre-development topography and hydrology for the assessment of tunnelling hazard (Birch *et al.* 1998);

• railway maintenance strategy (UK) – prioritization of repair and maintenance for coastal route.

In the application of engineering geomorphology, the emphasis must be on focusing the input on the engineering requirements.

Conclusions

An understanding of landforms and of the natural processes which have shaped, and are continuing to shape, the landscape is now recognized as an important element in project planning and design. Geomorphological surveys carried out at an early stage in project development can avoid abortive design work, and therefore wasted money, or worse, late abandonment.

In particular, the engineering geomorphological map provides:

(1) an early indication of adverse or hazardous processes;
(2) a better understanding of topographic, geological and geotechnical conditions in advance of site access;
(3) a conveniently scaled map for field reconnaissance and for communication of relevant information on environmental and archaeological aspects;
(4) a basis for the design of subsurface investigations;

(5) a basis for engineering geological evaluations; and
(6) an indication of possibly adverse environmental or knock-on effects relating to pipeline construction.

In essence, geomorphological surveys are an effective and low-cost tool which can streamline the planning and design of pipeline projects and other linear projects, by the early identification of potentially adverse ground conditions.

References

BIRCH, G. F. 1989. Applications of geomorphology to small hydro schemes. *Quarterly Journal of Engineering Geology* **22**, 231–239.

BIRCH, G. P 2001. Rapid evaluation of ground conditions for the 'Ave' railway, Spain. *This volume.*

BIRCH, G. P., LANCE, G. A. & HEWISON, L. R. 1998. Hazard assessment for the Athens Metro, Greece. *Proceedings of the International Conference on Urban Ground Engineering, Hong Kong*, November 98. Institution of Civil Engineers, London.

BRUNSDEN, D., DOORNKAMP, J. C., FOOKES, P. G., JONES, D. K. C. & KELLY, J. M. H. 1975. Large scale geomorphological mapping and highway engineering design. *Quarterly Journal of Engineering Geology*, **8**, 227–53.

HADLEY, E. 1991. Engineering a greener pipeline: a practical approach. *Pipes and Pipeline International*, Jan.–Feb.

Rapid evaluation of ground conditions for the AVE railway, Spain

G. P. Birch

Consultant Engineering Geologist, Sevenoaks, Kent, UK

Objectives

The development of high-speed railways across Europe, spurred on by the commencement of the Channel Tunnel in 1986, saw the development of Spain's first European gauge high-speed line, the Alta Velocida Espana (AVE), between Madrid and Sevilla, southern Spain, aided by European Economic Community funding (Fig. 1). The construction timetable required completion by April 1992, in time for inauguration by King Juan Carlos at the opening of Expo'92 in Sevilla.

Whilst the greater part of the 470 km new alignment from Madrid was advancing well, the final 125 km between Cordoba and Sevilla had to be squeezed into a $2\frac{1}{2}$ year design and construct period. This followed a late decision to switch from an upgrade of the existing route to a new alignment along the valley of the River Guadalquivir, one of Spain's five major rivers (Fig. 2).

A major landslip occurred during construction on the Cordoba–Sevilla section destroying both the new works and adjacent rural railway (Fig. 3). This prompted the Railway Inspectorate for RENFE, the Spanish National Rail Network, to commission an independent technical audit to address, in particular, the geotechnical aspects of the construction works.

Following the landslip, and a number of other very worrying incidents, the Inspector General expressed concern for the quality of construction and integrity of the earthworks, including in particular:

- the absence of a project-wide geological study and site investigation;
- settlements at embankment/structure transitions;
- erosion and minor failures on embankments;
- instability in cuttings;
- adequacy of cross-track drainage and effect on adjacent railway;
- adequacy of design of bridges and retaining walls;
- adequacy of quality control.

The client's perception, therefore, was that the entire 125 km between Cordoba and Sevilla should be subject to a complete geotechnical appraisal comprising a campaign of regularly spaced boreholes along the railway formation.

Regional setting

The route traverses the gently undulating flanks of the Rio Guadalquivir immediately south of the Sierra Morena which form the southern edge of the Spanish Mesata.

The strata encountered comprise largely loose or partly consolidated gravels, sands, silts and soft clay of Quaternary age which have accumulated by the gradual infilling of the Guadalquivir Depression (Vanney 1971). Older basement rocks of Cambrian Age are encountered in the region of Almodovar del Rio where the railway tunnels (300 m) beneath a castle built atop a granite intrusion. A further outcrop of basement rocks is exposed to the west in a deep cutting.

Between the alluvial gravels and the basement rocks are contrasting blue-grey marly deposits laid down in a quiet lacustrine regime which prevailed during Turtoneien Times. These marls, known as 'margas azules', are characterized by a high proportion of the swelling clay minerals of the smectite group and this characteristic has a very significant bearing on its geotechnical behaviour (Gonzalez & Galan 1986).

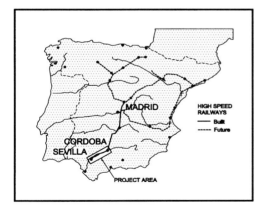

Fig. 1. Location map.

From: GRIFFITHS, J. S. (ed.) *Land Surface Evaluation for Engineering Practice*. Geological Society, London, Engineering Geology Special Publications, **18**, 83–89. 0267-9914/01/$15.00 © The Geological Society of London 2001.

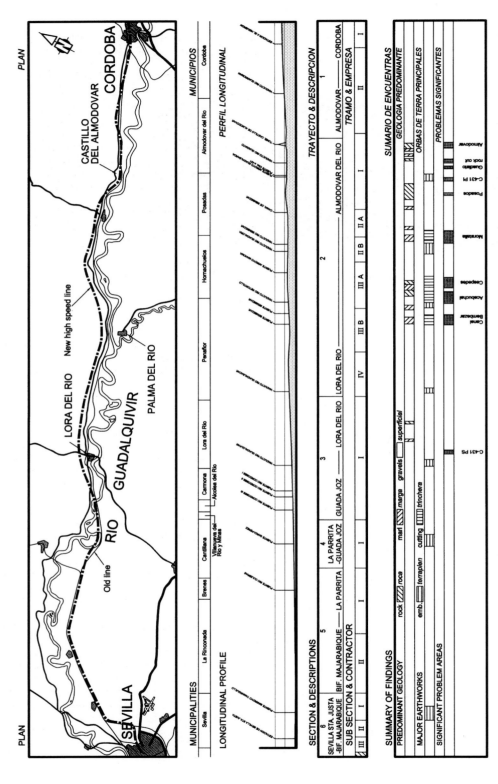

Fig. 2. Plan of the Sevilla to Cordoba section of the high-speed railway with longitudinal profile along the alignment and summary of the geology and earthworks.

Fig. 3. Landslide on the Cordoba–Sevilla section of the railway that occurred during construction.

Techniques

Following a rapid one-date route reconnaissance we were able to persuade the client that a more cost-effective and beneficial solution would be a phased approach in order to maximize the use of available information before commitment to expensive and inappropriate sub-surface investigations. The phased approach also provided the opportunity for feedback to the client at key stages of the programme.

Phase One: desk studies and field reconnaissance

The objective at this stage was to establish the geological/geomorphological setting to the project and the scope and quality of the geotechnical data generated during the design stage of each section of the works.

Sources of information included:

- published topographic and orthophoto maps at 1:25 000 scale (14 photomosaics);
- published geological maps at 1:50,000 scale and associated memoirs (six sheets);
- published technical papers from The Geological Society, Imperial College and University of Sevilla (13 references);
- unpublished technical reports and site investigation data from client and contractors (105 documents);

- stereo aerial photographs at 1:18 000 scale from national coverage;
- stereo aerial and oblique photographs taken for the project;
- meetings with representatives of the contractors.

A major component of the desk study was the characterization of geotechnical and geomorphological conditions through air photo interpretation and field reconnaissance to provide 'ground truthing'. This phase provided an early indication of the likely extent of additional ground investigations necessary to confirm or re-establish the geological conditions.

The field reconnaissance commenced with a two-day inspection (level 1 reconnaissance) using track-side access roads, which are a feature of the project, wherever possible.

Having confirmed the limited opportunity for geotechnical problems over those stretches of track at grade and on level ground, attention was focused for the remainder of the field reconnaissance period on a more detailed assessment on foot (level 2 reconnaissance). This survey covered the section of route between Almodovar del Rio in the east, and the Rio Guadalquivir crossing in the west.

Field record sheets designed to capture rapidly a wide range of factors pertinent to the geotechnical conditions, were completed for the entire route as well as for each major structure, including those falling within sections of track at grade (Fig. 4). The locations of each of the 148 field record sheets were indicated on the 1:50 000 scale route maps. Photographs were taken to illustrate specific features identified in the fieldwork.

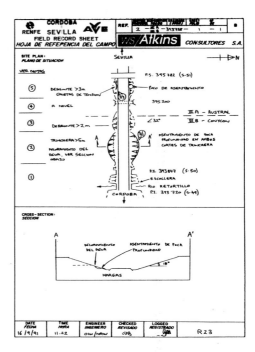

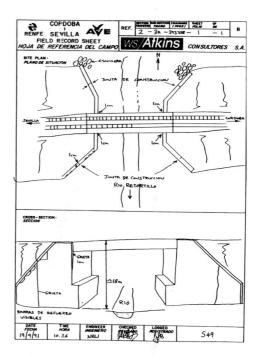

Fig. 4. Field record sheets.

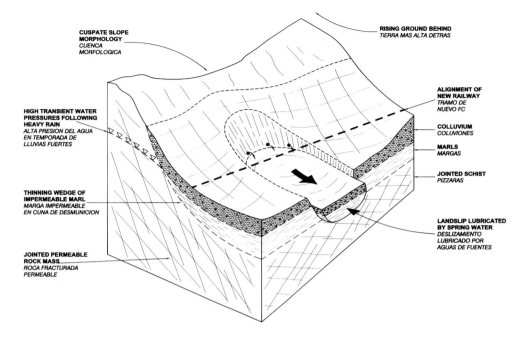

Fig. 5. Generic model of the landslide failures in the marl.

Of the more serious problems noted by the field reconnaissance, many involved the notoriously weak marls of the Guadalquivir Depression. The presence and indeed the significance of these marls had evidently not been recognized in the design of cuttings, and in some cases possibly foundations. The anticipation at outline design of cut slope of 1h to 1v (45°) would appear to have been over-ambitious in the light of construction experience. For the more serious problems, specific recommendations were made as to what further investigations were necessary in order to develop remedial designs.

The findings of Phase One were presented as an interim report accompanied by the field record sheets and 1:50 000 scale route maps highlighting those areas requiring special attention.

Phase Two: field investigations

The primary function of Phase Two was to elucidate and re-evaluate ground conditions at eight locations where specific geotechnical problems were indicated. The two most severe problem locations shared the same generic cause, namely triggering of incipient failures within the troublesome marls (Fig. 5).

Following discussions with the client, a limited campaign of subsurface investigations was carried out comprising 11 carefully targeted rotary cored boreholes at four of the eight locations of residual problems. In addition, general guidelines were given on the resolution of project-wide problems of a more generic nature relating to erosion and washout affecting both cuttings and embankments and, more especially, bridge abutments.

Results

The absence of a desk study for the 125 km route had left a gap in the design process such that inappropriate parameters were being adopted. This project-wide overview of the geological and geomorphological development of the region provided a basis for understanding of the engineering difficulties experienced by individual contracts and the underlying cause of the major landslip processes. For example, geomorphological mapping from 1:18 000 scale stereo aerial photographs and published 1:25 000 photomosaic maps revealed the location where the alignment traversed former stream courses and abandoned meanders of the Rio Guadalquivir (Fig. 6). It was, therefore, no coincidence to find severe

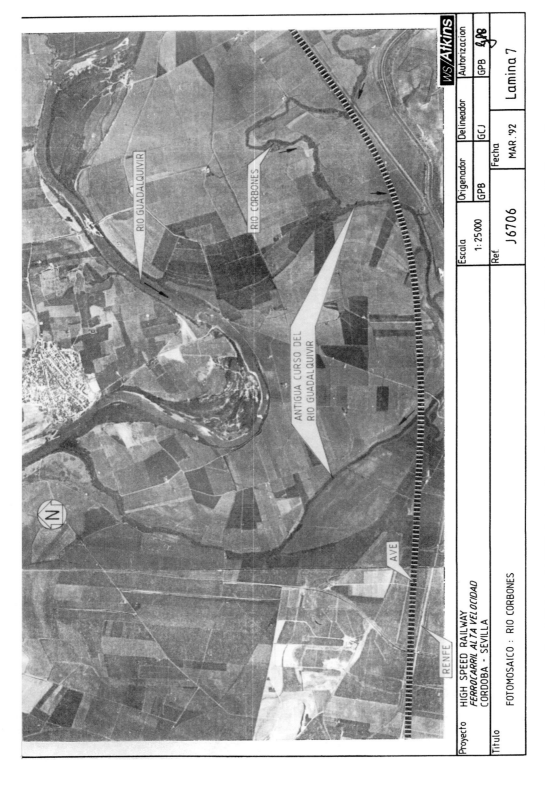

Fig. 6. Extract of 1:2500 photomosaic showing the former meanders of the Rio Guadalquivir traversed by the new line.

embankment settlements adjacent to piled bridge structures at these locations.

The overall conclusion was that the client welcomed a fresh, rapid and objective evaluation free from political, contractual or technical preconceptions. The end-product was a survey of the land surface that would be directly accessed for the design of any future remedial works.

References

GONZALEZ, I. & GALAN, E. 1986. Origin and environmental conditions of the Tertiary marine deposits of the Sevilla area, Guadalquivir basin, Spain. *Tenth Conference on Clay Mineralogy and Petrology, Ostrara*, 209–217.

VANNEY, D. L. 1971. *Le Bas Guadalquivir*. Ru Vase de Velazqiez.

Preliminary landslide hazard assessment in remote areas

J. H. Charman

Consultant Engineering Geologist, Milford, Guildford, Surrey, UK

Rationale

In remote areas decisions on the need for remedial and maintenance works on linear projects such as roads and canals may involve the evaluation of risk based on the identification and distribution of definable hazards. A terrain model provides the basis for a classification scheme to enable this to be done cost effectively. In addition it facilitates the identification of problem areas that may require more detailed subsequent design investigation. In the absence of more detailed investigation it provides the basis for a conceptual construction approach.

The site

This example relates to hill irrigation schemes in the Himalaya of Nepal and Bhutan and involved the upgrading of existing village canals irrigating between 50 and 200 hectares of agricultural land for rice production. Each canal ran across steep slopes and was experiencing difficulties with slope instability. In the feasibility stage a limited number of projects had to be selected from a large initial listing.

Area model

The Himalaya form part of an active fold mountain belt characterized by moderate to high rates of uplift, regular seismic events and extreme relief. They are also subject to a tropical monsoon climate, at least at lower elevations, which causes significant weathering and erosion.

Mountain-building activity is not uniform but comprises episodes of uplift interposed between quieter periods. In active times rivers rapidly downcut and form steep incised valleys. In quieter periods weathering and slope evolution and more mature rivers result in a less severe landscape. The model in Figure 1 uses this concept to classify the landscape.

The active lower slopes of Zone 4 represent the current phase of slope steepening caused by stream incision, while the overlying more gentle slopes of Zone 3 mark an earlier break in tectonic activity. Zone 5 is the area of deposition where alluvial terraces of varying age line the lower valley slopes and floor. A full description

of the model can be found in Fookes et al. (1985). Most hill canals in the study area run through Zones 3, 4 and 5.

The technical problems

Hill canals require an intake on a tributary channel above the main valley. They must run to a predetermined gradient to the irrigation area and thus inevitably cross the steep slopes of Zone 4 before emerging onto gentler Zone 3 slopes where the agricultural land is located. The major problem for these canals is to identify the existing landslides and to use methods of design and construction that do not exacerbate these, minimize new instability elsewhere and are locally sustainable.

Techniques used

Aerial photographs at a nominal scale of 1:25 000 were available in stereo pairs for interpretation, but the extreme relief sometimes limited stereoscopic viewing. These were used to carry out an initial land classification of each project area into mountain zones or units (Fookes et al. 1985). When possible, the units were further subdivided into land elements that were visually recognizable as a distinct surface type (Fig. 1). These units also represented a distinct approach to appropriate design and construction to sustain the life of the canal.

Other factors important to the occurrence of landslides and erosion were identified. These included: climate, particularly rainfall and temperature, both of which varied significantly with altitude; geology, particularly soil and rock type and rock structure; land use; and the location of other development projects in the area. Existing information on these factors was gathered from local sources.

Selected aerial photos were enlarged to provide 1:5000 scale base maps for walk-over survey. This allowed confirmation of the land elements and the addition of other elements that were not distinguished in the initial photo interpretation. During the survey the basic geology and landforms were mapped, the land use was differentiated, and surface and groundwater conditions were noted. Areas of existing instability were demarcated and the probable mechanism of failure noted (Varnes 1978).

From: GRIFFITHS, J. S. (ed.) *Land Surface Evaluation for Engineering Practice*. Geological Society, London, Engineering Geology Special Publications, **18**, 91–95. 0267-9914/01/$15.00 © The Geological Society of London 2001.

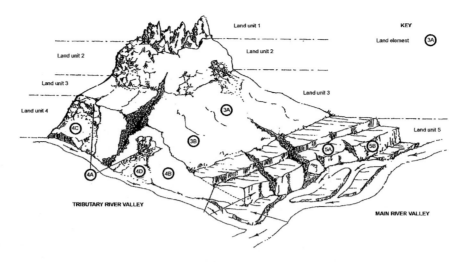

No	Description	No	Description
1	High altitude glacial and periglacial areas subject to glacial erosion, mechanical weathering, rock and snow instability and solifluction movements with thin rocky soil, boulder fields, glaciers, bare rock slopes, talus development and debris fans		
2	Free rock face and associated steep debris slopes subject to chemical and mechanical weathering, mass movement, talus creep, freeze-thaw, and debris fan accumulation.		
3	Degraded middle slopes and ancient valley floors forming shallow erosional surfaces subject to chemical weathering, soil creep, sheetflow, rill and gully development and stream incision.,	3A	Ancient erosional terraces covered with a weathered residual soil mantle generally up to 3m thick. Slope angle generally < 35° and stable. Often farmer terraced. Highly susceptible to water erosion
		3B	Degraded colluvium comprising landslide debris of gravel, cobbles and boulders in a matrix of silt and clay. Slope angle < 35°. Relatively stable. Often farmer terraced. Variable permeability
4	Steep active lower slopes with chemical and mechanical weathering, large-scale mass movement, gullying, undercutting at base and accumulation of debris fans and flows of marginal stability	4A	Bare rock slopes. Steep slope angles > 60°. Stability dependent on orientation of discontinuities, such as joints and bedding planes.
		4B	Rock slopes with mantle of residual soil usually < 2m thick. Steep slope angles > 45°. Prone to extensive shallow debris slides. Deeper instability as for 4A.
		4C	Active colluvium. Thick landslide debris often at base of slope and subject to active river erosion. Slope angle > 35°. Highly unstable, particularly during wet season.
		4D	Degraded colluvium. Thick landslide debris. Slope angle < 35°. Marginally stable and susceptible to gradual downslope creep during wet season
5	Valley floors associated with fast flowing, sediment laden rivers, and populated by sequences of river terraces.	5A	Top of old alluvial terraces above present river level. Generally flat to shallow, < 10°. Coarse granular and permeable soils. May be covered by a less permeable residual soil mantle.
		5B	Front scarp face of old alluvial terraces. Steep slope angle > 65o, but subject to sudden collapse when cementation breaks down under weathering or when subject to toe erosion.

Fig. 1. A mountain classification (after Fookes *et al.* 1985).

Local ground models

Completion of the desk study and field survey provided the necessary data to enable a landslide hazard assessment to be compiled. This was based on a pro forma which introduced a scoring rating for each of the identified causative factors (Fig. 2). A full description is provided in UNDP/ILO (1993).

Terrain

The basic terrain model was enhanced in this study to enable individual elements to be identified. This was important to provide a basis for allocating a relative hazard score. Each element was identified from field observations and defined the conceptual engineering approach.

Zone 3 is comparatively stable and scores 1. The steep slopes of Zone 4 can be divided into several elements.

Element 4B represents the undisturbed slope comprising a bedrock thinly covered by a weathering mantle of residual soil. This has a high potential for debris slides and scores 4. Once failure has occurred the relatively fresh rock of the back scarp (Element 4A) provides a firm foundation for the canal and scores 2. The slipped debris either continues to be seasonally active (Element 4C) scoring 4 or reaches a stable angle of repose (Element 4D) scoring 3. Zone 5 is divided into the flat terrace surfaces (Element 5A) scoring 1 and the steep river-cut faces prone to sudden slumps (Element 5B) scoring 4.

Geology

The scoring system is applied to rock type on the basis of its relative resistance to weathering. For example.

PROJECT: KURGHA
Sheet No: 1 of 4

Completed by: RG
Date; 11-2-92

FACTOR		SCORE	CHAINAGE					
			0–150	151–750	751–850	851–1200	1201–1250	1251–2100
TERRAIN CLASS'N	Land Element 3	1						
	Land Element 4A	2		2		2		
	Land Element 4B	4						4
	Land Element 4C	4			4			
	Land Element 4D	3	3				3	
	Land Element 5A	1						
	Land Element 5B	4						
GEOLOGY 1 Rock Type	Quartzite, Marble	1						
	Gneiss, Sandstone	2		2		2		
	Limestone	3						
	Phyllite	4						4
	Mica Schist	4						
GEOLOGY 2 Soil Type	Coarse Granular (gravel)	1						
	Fine Granular (sand,silt)	3	3		3		3	
	Cohesive (clay)	2						
GEOLOGY 3 Structure	Dip out of slope	4				4		4
	Dip into slope	2		2				
CLIMATE	Sub-alpine (3000–4500m)	1						
	Cool temperate (2000–3000m)	2	2	2	2	2	2	2
	Warm temperate (1200–2000m)	3						
	Sub-tropical (0–1200m)	4						
LAND USE	Dense forest	1						1
	Scrub/grass	2	2	2		2		
	Dry cultivation (khet)	2					2	
	Wet cultivation (paddy)	4			4			
	Fallow	3						
GROUND WATER	Dry	1	1	1		1		1
	Seepage	2					2	
	Moderate flow	3						
	Heavy flow	4			4			
HAZARD RATING			11	11	(17)	13	12	16

Fig. 2. Terrain hazard assessment pro forma.

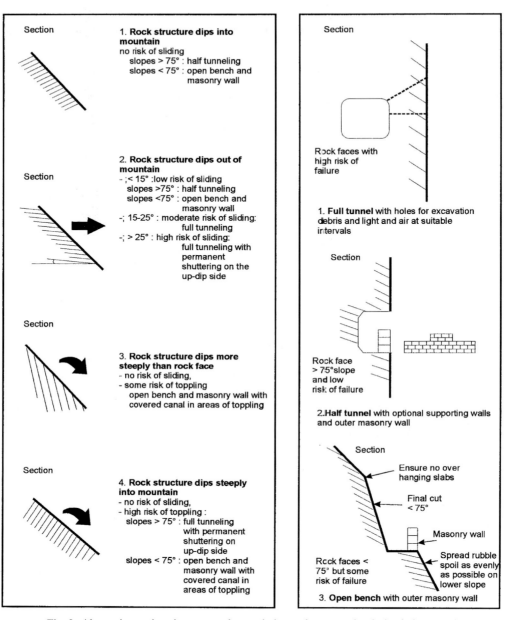

Fig. 3. Alternative engineering approaches to design and construction in land element 4A.

a resistant quartzite scores 1 while a mica schist, highly susceptible to weathering, scores 4. For soil type a coarse granular soil scores 1 while a weaker cohesive soil scores 2. A structure in which the main discontinuity dips out of the slope and provides a failure plane scores 4 and a dip into the slope scores 2.

Climate

Climate is heavily influenced by elevation and changes dramatically over short distances. A tropical monsoon climate combining high temperature and high seasonal rainfall is the most destructive and induces the highest rate of weathering: this scores 4. Conversely, a cool temperate climate at elevations of between 2000 m and 3000 m scores 2.

Land use

In an area where farmers strive to terrace the slopes and irrigate them profusely to generate rice paddy, land use has a significant effect on stability. Disturbance of this vegetation cover removes the binding effect of roots, the protection against rainsplash and rill erosion and allows greater infiltration of water. Excess irrigation water is often poorly managed. Undisturbed dense forest scores 1 while wet paddy cultivation scores 4.

Groundwater

The level of groundwater in the slope and the excess pore pressure that this implies directly influences potential slope instability. Qualitative assessment allows differentiation between saturated conditions which score 4 through to dry conditions which score 1.

Use of the ground models

In remote areas where access is limited, detailed ground investigation may be impossible. Decisions on project feasibility and the formulation of the detailed design may have to be made on the basis of desk study and walk- or drive-over survey alone. Construction methods are devised for implementation by a local workforce and local materials are utilized.

In these circumstances terrain classification is a particularly important part of the decision-making process. It provides a semi-quantitative and rapid method of appraisal, and identifies the potential problem areas for detailed geomorphological and engineering geological mapping. This allows conceptual alternative designs to be developed so that they can be incorporated when construction reveals the actual site conditions.

The hazard assessment described here could be carried out by local engineering staff under the guidance/ supervision of a qualified engineering geomorphologist/ geologist. The method has not been mathematically tested and is open to improvement but it has been employed and field tested with some success to classify road and irrigation canal sections in the Himalaya (UNDP/ILO 1993).

The subdivision of the general terrain model into elements provides the basis for an appropriate engineering approach to remediation and design. For example, land element 4A is indicative of relatively fresh rock in the back scarp. It is therefore a potential source of rock for stone and aggregate, but may be difficult to excavate. The approach to canal construction in this terrain element is illustrated by the conceptual construction outline given in Figure 3.

The level of engineering achievement attained on the basis of an approach such as this is dependent on supervision by an experienced and practical construction professional, the ability to adapt to conditions as they are revealed, and the use of local materials and skills which allow sustainable maintenance. Further reading on general design and construction measures is contained in Fookes *et al.* (1985). Some examples of successful implementation are given in UNDP/ILO (1993).

References

FOOKES, P. G., SWEENEY, M., MANBY, C. N. D. & MARTIN, R. P. 1985. Geological and geotechnical engineering aspects of low-cost roads in mountainous terrain. *Engineering Geology*, **21**, 1–152.

UNDP/ILO. 1993. *A manual for environmental protection measures for hill irrigation schemes in Nepal*. Nepal SPWP Manual No. 1 International Labour Organisation, Geneva.

VARNES, D. J. 1978. Slope movement and types and processes. *In*: SCHUSTER, R. L. & KRIZEK, R. J. (eds) *Landslides: Analysis and Control*. Transport Research Board, National Research Council, USA, Special Report 176, Ch. 2.

Subsidence hazard in Berkshire in areas underlain by chalk karst

C. N. Edmonds

Peter Brett Associates, Reading, Berkshire, UK

Purpose of survey

During the last ten years a number of ground subsidence events have occurred in the northwest part of Reading. Many of the subsidence events resulted in structural damage to existing properties (see Plates 1 and 2). On the basis of the properties inspected to date it appears that the local housing has been constructed mostly upon conventional strip footings bearing onto naturally occurring soils. The increasing number of recorded subsidence events is of concern to planners, developers and insurers. Consequently the aim of the survey was to identify the nature and extent of subsidence hazard in the local area.

The site

The northwest part of Reading, generally referred to as Caversham, is shown in Figure 1. It largely comprises a south to southeasterly dipping land surface, overlooking the River Thames. North of the Thames the land surface is dissected by a NNW–SSE trending valley feature known as Hemdean Bottom. This divides the westerly Caversham Heights area from the easterly Caversham Park and Emmer Green areas. The Thames lies at just below 40 m AOD and northwards the land rises to above 80 m AOD. The floor of Hemdean Bottom generally lies between 40 m and 50 m AOD.

Geology

The published geological map at 1:10 560 scale (British Geological Survey County Series Berkshire Sheet 29 SE) for this area shows the entire district to be underlain by Cretaceous Upper Chalk, overlain by a Tertiary Reading Beds outlier to the northeast side of Hemdean Bottom. The former London Clay Formation cover has been eroded away within the study area. The chalk is a very weak to moderately strong, white, porous, soluble, carbonate rock with flint bands. The chalk surface is highly weathered where exposed and frost shattered. The Reading Beds are now referred to as the Reading Formation, part of the Palaeogene Lambeth Group.

Locally the Reading Formation comprises mostly clays in the upper portion of the stratum overlying mostly sands in the lower portion. The higher ground is capped by Hill Gravel, a sandy gravel with some silt and clay horizons. Valley Loam is present along the floor of Hemdean Bottom extending southwards towards the Thames where it joins with the Alluvium that occupies the Thames floodplain. The Valley Loam is a calcareous silty sandy clay with occasional gravel, while the Alluvium consists mostly of silt with occasional organic horizons (Blake 1903).

A simplified geological plan is shown in Figure 2. The Hill Gravel is now thought to be a Post-Anglian fluvial terrace deposit laid down by the proto-Thames when the river bed lay at a higher topographical level (Jones 1981). The Valley Loam appears to be a brickearth/head deposit.

Geomorphology

Tectonic downwarping of the London Basin in the early Tertiary resulted in a number of transgressions which laid down a sequence of Palaeogene strata across the Thames Valley area containing Reading. Continuing tectonic activity through into the Neogene, and erosion, resulted in the stripping back of the Palaeogene cover (Jones 1981). Following removal of the London Clay Formation cover across the area and a fall in water table level below the sub-Palaeogene surface, the potential has existed for downward infiltration of groundwater through the Reading Formation to initiate dissolution of the Chalk below. On the basis of erosional evidence (e.g. unroofing of the Weald, Jones 1981) it seems likely that these conditions have been present from late Tertiary to early Pleistocene times onwards.

During the Pleistocene the Anglian ice front advanced as far south as the Vale of St Albans. Reading lay beyond the ice front and was subjected to periglacial weathering (Jones 1981). Another important Pleistocene event was the diversion of the proto-Thames river by the Anglian ice. Both before and after impedance of the proto-Thames river, fluvial and fluvio-glacial deposits have been laid down along its ancestral and modern

From: GRIFFITHS, J. S. (ed.) *Land Surface Evaluation for Engineering Practice*. Geological Society, London, Engineering Geology Special Publications, **18**, 97–106. 0267-9914/01/$15.00 © The Geological Society of London 2001.

Plate 1. Internal view of structural cracking damage to a property at Buxton Avenue, Reading following subsidence over a solution feature.

Plate 2. External view of compaction grouting stabilization works being carried out at the same property at Buxton Avenue, Reading.

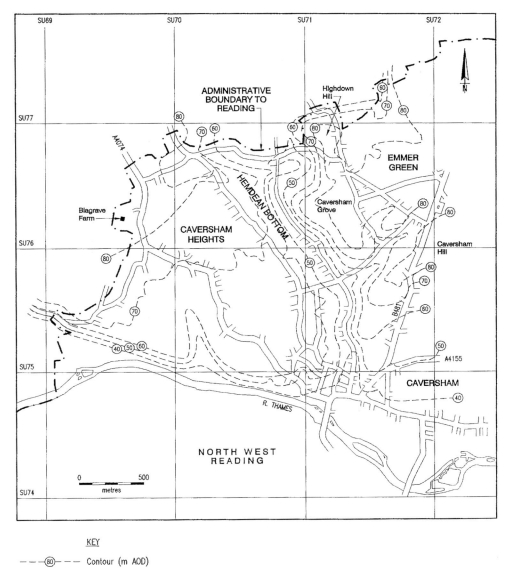

KEY

— — —⑧⑩— — —   Contour (m AOD)

Fig. 1. Study area location and topography.

course (Gibbard 1977, 1983). These events were mainly focused on the Vale of St Albans to the east, but they also influenced deposition of the Hill Gravel in the Reading area.

Solution feature formation results from the progressive downward infiltration of groundwater (undersaturated with respect to calcium carbonate) into the Chalk surface, where it dissolves the rock gradually over a long period of time (>10 000 years). The process produces solution features that penetrate downwards

along joint fractures. As a result, solution features are preferentially developed at the interface of the Chalk surface with the overlying cover deposit. This is discussed by West & Dumbleton (1974), Rigby-Jones *et al.* (1993) and McDowell & Poulsom (1996), and is illustrated schematically in Figure 3. The range of solution features formed includes solution pipes, swallow holes and sinkholes (Edmonds 1983).

In northwest Reading it is suggested that solution features have formed below the Reading Beds and Hill

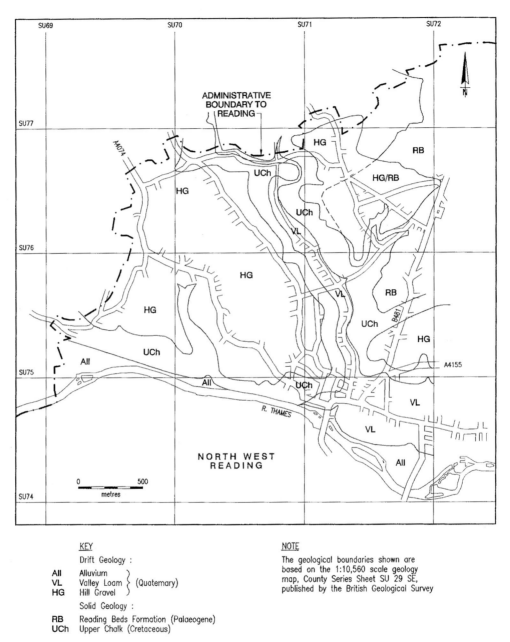

Fig. 2. Study area simplified geology.

Gravel at various times in response to climatic change, depositional and erosional events. The most active solution feature formation phases were probably associated with the seasonal thaw of periglacial frozen ground (Higginbottom & Fookes 1970). It is suggested that many of the solution features present today possibly date from Anglian times onwards. It is expected that within the study area, solution features will be associated with the Chalk/Reading Formation and Chalk/Hill Gravel interfaces. Slightly acidic groundwater collecting upon these cover deposits will have had ample opportunity to form concentrated flows able to percolate

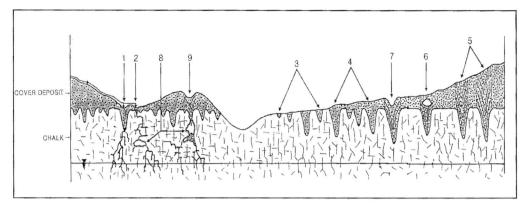

Fig. 3. Schematic diagram to illustrate range of natural solution features present on the Chalk (after Edmonds 1987). Solution feature types: swallow holes (1), (2); erosional remnants of solution pipes (3); solution pipes (4), (5); solution pipe with upward migrating air-filled void (6); subsidence sinkhole over a solution pipe (7); solution cavities connecting with solution-widened joints (8); collapse sinkhole over a solution cavity (9).

downwards into the chalk surface over a long period of time to produce solution features. This is shown schematically in Figure 4.

Techniques used and limitations

Broad-scale geomorphological mapping has been carried out within the study area to determine the terrain and surface/subsurface drainage characteristics. The mapping technique was applied following the principles of subsidence hazard mapping developed by Edmonds (1987, 2001). The mapped boundaries of terrain units have not been set out by accurate survey, but have relied on topographic base maps of various scales and observation by walkover survey. The extensive coverage of existing development has also influenced the accuracy of mapping.

A review of historic maps has revealed a number of old chalk quarries, clay pits and gravel workings within the study area. Their locations were noted during the mapping survey and their influence upon the local landform assessed. However, none of the man-made workings were particularly extensive, obliterating the natural land profiles. Therefore man's historical impact on the local terrain is considered not to have greatly affected the survey results.

Study area morphology

The morphology of the northwest Reading study area is shown in Figure 5. The land surface comprises a series of valleys with tributaries that drain southwards towards the River Thames.

There is a deep-cut minor valley that extends in a NNW–SSE orientation on the west side of the Caversham Heights area. Further east is the NNW–SSE trending Hemdean Bottom valley, the floor of which is relatively narrow, less than 100 m wide in many places. Mid-way along the east side of the Hemdean valley is a minor tributary valley that extends northeastwards, dividing into two parts at the head of the valley. Another minor tributary valley extends westwards into the Caversham Heights area a little further north along the Hemdean valley.

The land surface profile of the interfluves consists of a series of stepped surfaces that lead upwards from the valley sides to a number of local topographic 'highs'. To the west of Hemdean Bottom the stepped surfaces climb up to a hill brow trending ENE from Blagrave Farm (see Fig. 1). On the east side of Hemdean the high points comprise Caversham Hill and another area to the north centred on Highdown Hill and Caversham Grove. The stepped surfaces generally take the form of flat to slightly inclined (1 to 2°) bench levels which are bounded by breaks of slope. On the uphill side is a concave break of slope, while on the downhill side is a convex break of slope. Within the study area it appears that generally slope angles do not exceed 5° between successive steps; however, locally steep slope angles (5 to 10° or more) are present where the interfluves finally step down into the valley floors of the Thames and Hemdean Bottom.

The conceptual model

Ground subsidence over solution features may be triggered in many ways, e.g. by heavy rainfall, water flows

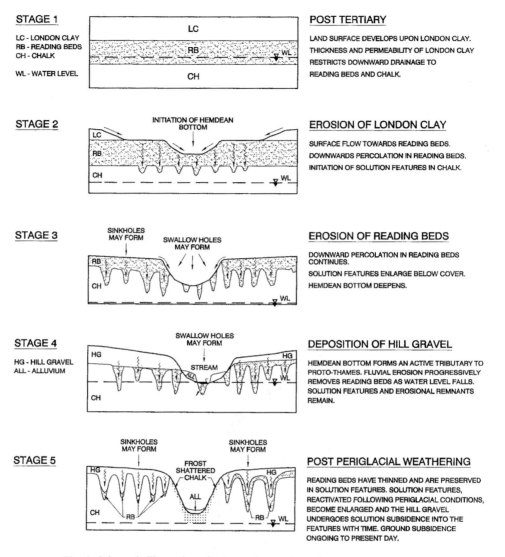

Fig. 4. Schematic illustration of solution feature formation in northwest Reading.

from soakaways, garden watering, leaking drains and others. As the deposits which infill a solution feature undergo solution subsidence they become weakened, loosened and disturbed. When destabilized by any of the above subsidence triggers, available subsurface void space may be rapidly transmitted upwards through the infill to reach the surface and form a sinkhole. This is caused by breakdown of metastable voids, internal piping, erosion and wash-through of fines within the infill. These mechanisms are widely reported by West & Dumbleton (1972), Edmonds (1988), McDowell (1989), McDowell & Poulsom (1996) and Rhodes & Mary-

church (1998). The instability that leads to subsidence problems is therefore associated with the destabilization of infilling deposits within pre-existing solution features and not the formation of new features. This is further illustrated by Figure 3. In the case of very large, mature features that are interconnected with solution-widened joints in the Chalk at depth, it is also possible, by erosion at the base due to flowing water, for additional upward-migrating voids to be introduced to cause subsidence at the surface.

In northwest Reading, as illustrated by Figure 4 it is envisaged that solution features have been formed below

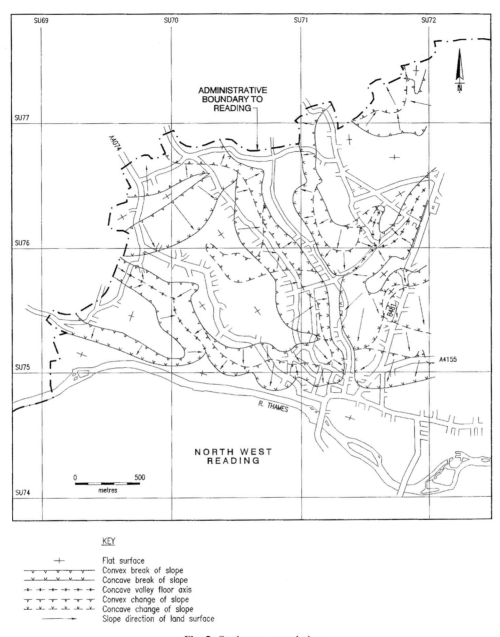

Fig. 5. Study area morphology.

both the Reading Beds and the more recent Hill Gravel cover deposits. An earlier phase of solution feature development is thought to have occurred by downward infiltration of water through the sub-Palaeogene surface (Stages 2 and 3, Fig. 4). A later phase of solution feature development is considered to have taken place at the Hill Gravel/Chalk interface. Fluvial erosion and the deposition of the Hill Gravel has contributed to the erosion of the Reading Beds (Stage 4, Fig. 4). Consequently, where the sub-Palaeogene surface has been destroyed, solution features will also have been removed. However, locally in northwest Reading the difference in

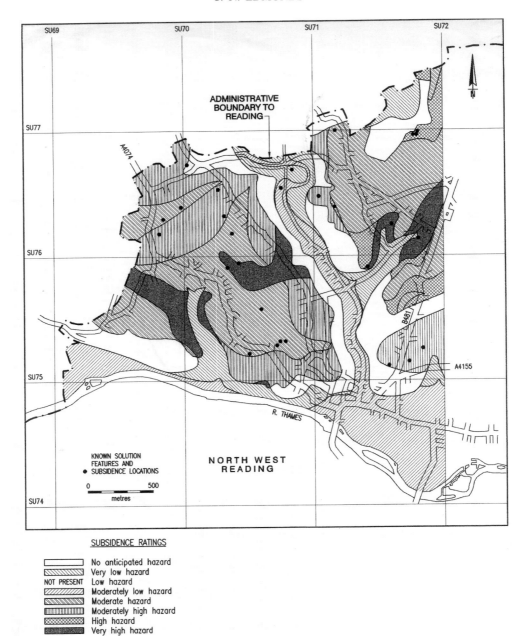

Fig. 6. Subsidence hazard map.

level between the sub-Palaeogene surface and the sub-Hill Gravel surface appears to be only about 10 m or so. It is not unusual for solution pipes to extend more than 10 m in depth, and 20 m or so is quite common. Therefore, it seems that total destruction of the solution features associated with the sub-Palaeogene surface probably did not occur. Hence, erosional remnants of solution pipes infilled with Reading Beds could have been preserved below the Hill Gravel (Stage 4, Fig. 4). In later favourable, post-periglacial climatic circumstances, permitting resumption of the downward infiltration of water into the Chalk, it is proposed that the erosional pipe remnants became a focus for renewed dissolution activity (Stage 5, Fig. 4). It is believed that the cumulative effect of this solution feature formation model begins to explain why solution features are prevalent in northwest Reading and why there is a continuing ground subsidence problem.

Although at this stage the database of known subsidence events is limited, a subsidence hazard mapping exercise has been carried out, using the techniques explained elsewhere, to attempt to better define the likely occurrence of future subsidence events. The subsidence hazard mapping techniques are explained in detail in Edmonds (2001). The surface morphology, subsurface geology, geomorphological development and the hydrogeology of the study area have been utilized to prepare a subsidence hazard map. The hazard map is shown in Figure 6.

Results of the hazard mapping

During the original development of the hazard mapping techniques, three study areas were chosen, each $100 \, km^2$ in size. Within the study areas the number of solution features was greatly enhanced by the interpretation of aerial photographs (viewed stereoscopically) and airborne multispectral scanner imagery. The larger numbers of features recorded increased confidence that representative cavity occurrence spatial patterns were being analysed and that locational bias was minimized.

When applying the techniques to a smaller urban study area, certain difficulties arise. Firstly it is often the case that the numbers of solution features and subsidence problems recorded are relatively small. Secondly locational bias is evident because of the nature of the surface development. Consequently the number of features and subsidence problems recorded may not, on a local area basis, fully reflect the predictions of the hazard mapping techniques which have been developed from a much larger database.

The original research database (Edmonds 1987), which was used to develop the hazard mapping technique, comprised 2226 solution features. When each of the solution feature locations in the database was subjected to hazard assessment and compared with the hazard categories, it was found that the number of features per hazard category tended to increase with ascending hazard rating as shown in Table 1. Within the Reading study area the relationship between hazard rating and solution feature occurrence is also shown in Table 1 for comparison.

The research database suggests that the increase in solution feature occurrence and related subsidence activity tends to be associated with hazard rating of 'Moderate' and above. In the Reading study area, comparison of solution features and subsidence occurrence with hazard rating reflects the general findings of the previous research, in locating those areas where ground subsidence is most likely to occur. Again, the areas of ground which are prone to solution feature occurrence and subsidence are mostly rated as 'Moderate' and above.

In the Reading study area the relationship between absolute numbers of recorded features, subsidence activity and increasing hazard ratings does not follow quite the same pattern as exhibited by the original research database (Table 1). Factors such as study area size, restricted solution feature data and locational bias explain this in part. In addition, the proportional spatial distribution of the various hazard zones will also influence matters. It is notable that this area lies south of one of the three $100 \, km^2$ study areas used as part of the original research, mentioned above, where a larger representative set of data more closely reflected the predicted pattern of solution feature occurrence beyond the urban fringe of Reading.

In addition, within the section on 'The conceptual model' above, attention was drawn to the relative proximity of the sub-Hill Gravel and sub-Palaeogene surfaces in the study area. It was suggested that where the gravels overlay solution pipe remnants infilled with Reading Beds (Fig. 4), they could act as a focus for

Table 1. *Subsidence hazard classification*

Subsidence hazard rating	Numerical range of each hazard rating category	Natural cavity occurrence (%) based on research database	Natural cavity occurrence based on Reading study area	
			(%)	*(no.)*
No anticipated hazard	<55	0.1	0	0
Very low hazard	55–89	0.5	3	1
Low hazard	90–136	3	Not present	
Moderately low hazard	137–200	4	3	
Moderate hazard	201–300	7	32	9
Moderately high hazard	301–400	12	29	8
High hazard	401–600	23	14	4
Very high hazard	>600	50.4	19	5

reactivation of solution feature development beneath the gravels. Consequently below the gravels, particularly within the 'Moderate Hazard' rated zones, there may be a higher than average density of solution features liable to lead to subsidence because of the favourable proximity effects of the two dissolution weathering interface levels.

Applications and uses

The hazard mapping techniques have wide application to the Chalk outcrop. They can be used to determine areas prone to subsidence, as illustrated by the above example, to assist planners, developers and insurers. Planners can then make the identification and mitigation of subsidence a condition of planning in line with PPG 14 *Development on unstable land* (Department of the Environment 1989) and the new Annex 2: Subsidence and Planning (Department of the Environment Transport and the Regions 2000). New-build developers and their geotechnical consultants can be alerted to the hazards and take appropriate precautions to ensure that suitably designed foundations are installed. Insurers can review the existing and future potential for subsidence associated with areas such as northwest Reading, and the hazard map could be used as a basis for decision-making when considering their insurance liability for this form of geohazard.

A further use for the technique is to understand the contamination susceptibility of the Chalk aquifer. Solution features can form preferential pathways for contaminants to pass rapidly down into the aquifer without attenuation. Hence it may be used to evaluate and check aquifer protection zoning around water supply boreholes, where it is normally difficult to take account of karst development by conventional hydrogeological modelling. The mapping might also be used as a basis to assess the possibility of environmental impact of certain potentially contaminating activities (e.g. landfills, petrol filling stations, engineering works, etc.) upon the Chalk aquifer. Environmental risk is greatly enhanced where solution features are present.

References

BLAKE, J. H. 1903. *The geology of the country around Reading.* Sheet 268, Memoirs of the Geological Survey, HMSO, London.

DEPARTMENT OF THE ENVIRONMENT. 1989. *Planning Policy Guidance: Development on unstable land.* PPG 14, HMSO, London.

DEPARTMENT OF THE ENVIRONMENT, TRANSPORT AND THE REGIONS. 2000. Planning Policy Guidance: Development on unstable land Annex 2: Subsidence and Planning. Consultation Paper **PPG14**, HMSO, London.

EDMONDS, C. N. 1983. Towards the prediction of subsidence risk upon the Chalk outcrop. *Quarterly Journal of Engineering Geology,* **16,** 261–266.

EDMONDS, C. N. 1987. *The engineering geomorphology of karst development and the prediction of subsidence risk upon the Chalk outcrop in England.* PhD Thesis, University of London.

EDMONDS, C. N. 1988. Induced subsurface movements associated with the presence of natural and artificial underground openings in areas underlain by Cretaceous Chalk. *In*: BELL, F. G., CULSHAW, M. G., CRIPPS, J. C. & LOVELL, M. A. (eds) *Engineering Geology of Underground Movements.* Geological Society, London, Engineering Geology Special Publications, **5,** 205–214.

EDMONDS, C. N. 2001. Predicting natural cavities in Chalk. *This volume.*

EDMONDS, C. N., GREEN, C. P. & HIGGINBOTTOM, I. E. 1987. Subsidence hazard prediction for limestone terrains, as applied to the English Cretaceous Chalk. *In*: CULSHAW, M. G., BELL, F. G., CRIPPS, J. C. & O'HARA, M. (eds) *Planning and Engineering Geology.* Geological Society, London, Engineering Geology Special Publications, **4,** 283–293.

GIBBARD, P. L. 1977. The Pleistocene history of the Vale of St Albans. *Philosophical Transactions of the Royal Society of London,* **B280,** 445–483.

GIBBARD, P. L. 1983. Slade Oak Lane. *In*: *Quaternary Research Association Field Guide for Annual Meeting, Hoddesdon, 'The Diversion of the Thames'*, 85–91.

HIGGINBOTTOM, I. E. & FOOKES, P. G. 1970. Engineering aspects of periglacial features in Britain. *Quarterly Journal of Engineering Geology,* **3,** 85–117.

JONES, D. K. C. 1981. *Southeast and Southern England.* University Paperbacks, Methuen.

McDOWELL, P. W. 1989. Ground subsidence associated with doline formation in chalk areas of southern England. *In*: BECK, B. F. (ed.) *Engineering and Environmental Impacts of Sinkholes and Karst.* Proceedings of the Third Multidisciplinary Conference on Sinkholes, St Petersburg Beach, Florida. Balkema, Rotterdam, 247–255.

McDOWELL, P. W. & POULSOM, A. J. 1996. Ground subsidence related to dissolution of Chalk in Southern England. *Ground Engineering,* March, 29–33.

RHODES, S. J. & MARYCHURCH, I. M. 1998. Chalk solution features at three sites in southeast England: their formation and treatment. *In*: MAUND, J. G. & EDDLESTON, M. (eds) *Geohazards in Engineering Geology.* Geological Society, London, Engineering Geology Special Publication No. 15, 277–289.

RIGBY-JONES, J., CLAYTON, C. R. I. & MATTHEWS, M. C. 1993. Dissolution features in the Chalk: from hazard to risk. *In*: INSTITUTION OF CIVIL ENGINEERS (eds) *Risk and Reliability in Ground Engineering.* Thomas Telford, London, 87–99.

WEST, G. & DUMBLETON, M. J. 1972. Some observations on Swallow Holes and Mines in the Chalk. *Quarterly Journal of Engineering Geology,* **5,** 171–177.

Creation of functional ground models in an urban area

R. J. G. Edwards

Earth Science Partnership, Leatherhead, Surrey, UK

Rationale

The ability to see the ground surface in an urban environment is limited by access and the amount of natural ground exposure. This is typically less than 10% by area. The assessment of the probable ground structure and the behaviour of critical ground elements can be achieved by comparison with non-urban environments exhibiting similar landforms. This model can be verified by a detailed study of historic maps and a review of existing local subsurface information.

The site

The City of Cardiff has largely been constructed in the last 150 years. The area studied relates to the whole city land area lying to the south of the A48T and between the west banks of the rivers Ely and Rhymney.

Area model

The area lies substantially on the Severn Levels, which comprise a recent tidal coastal margin zone of the Severn estuary. The area consists of a glaciated land surface comprising Mercia Mudstone and Dolomitic Conglomerate draped by a thin and partly discontinuous layer of glacial detritus that includes two identifiable layers of till and a consistent layer of fluvio-glacial sands and gravels which forms the dominant aquifer. These strata reflect Devensian and Flandrian glacial advance and recession.

To the south of the historic natural shoreline a series of post-glacial foreshore alluvia are superimposed on the late-glacial surface and form a confining layer to the underlying fluvio-glacial sand and gravel aquifer. These alluvia represent a series of more recent marine transgressions modified by sea defence construction works of Bronze Age to late mediaeval antiquity. Both the 'solid' and 'drift' geology is substantially masked by a variable thickness of waste and engineering fills of Recent anthropogenic origin (almost exclusively post-1800).

The glacial and pre-glacial geology is only exposed at the western and northern city margins. This sequence, with the exception of the anthropogenic fills, is mirrored in other identifiable areas in close proximity to Cardiff.

The technical problem

It was proposed to construct a barrage across Cardiff Bay to generate an attractive water frontage as an integral part of Cardiff Bay Development Corporation's inner city regeneration programme that included the derelict 'Tiger Bay' area of Cardiff Docks. The impact of this structure on the behaviour of the existing groundwater regime was recognized to represent a serious potential risk to the city's infrastructure. Also, as a result of a history of uncontrolled surface flooding within the city, the potential increase in groundwater level implied by the construction of the barrage was regarded as a matter of serious public concern.

In addition, the project was required by law to be promoted by a Parliamentary Bill necessitating that it be forensically examined in a series of Parliamentary Committee hearings at which objectors to the scheme were able to present their concerns and risk assessments.

A rational ground model was therefore an essential element in assessing the occurrence and behaviour of groundwater both before and after construction of the barrage. It was also required in order to generate qualitative and quantitative risk criteria and as a vehicle for critical discussion. No such model existed and little or no information regarding the distribution of strata and groundwater had been published or verified except for Anderson & Blundell (1965).

Techniques used

Comparable ground models of the basic 'solid' and 'drift' geology had been developed for adjacent areas of the Severn Levels using a combination of aerial photographic interpretation, geomorphological mapping, ground survey and the review of historical maps and other archival data. The results of these studies were published

From: GRIFFITHS, J. S. (ed.) *Land Surface Evaluation for Engineering Practice*. Geological Society, London, Engineering Geology Special Publications, **18**, 107–113. 0267-9914/01/$15.00 © The Geological Society of London 2001.

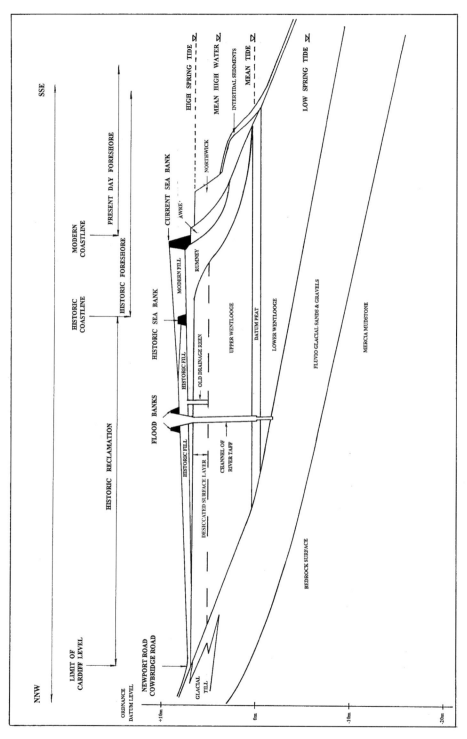

Fig. 1. Diagrammatic section through south Cardiff.

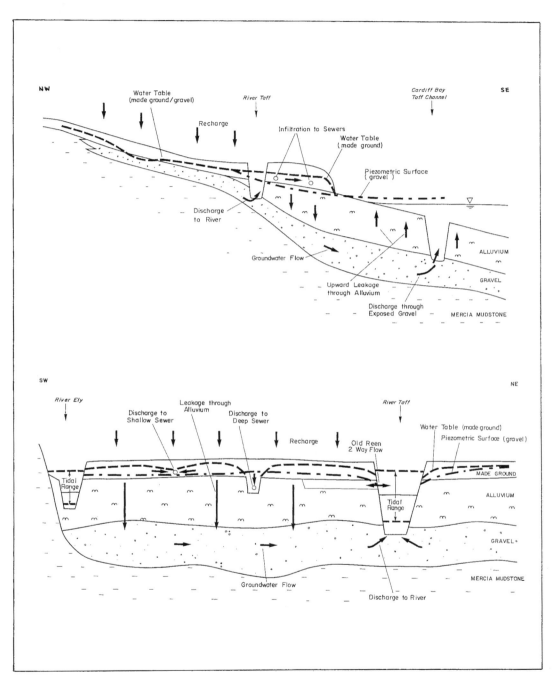

Fig. 2. Conceptual model.

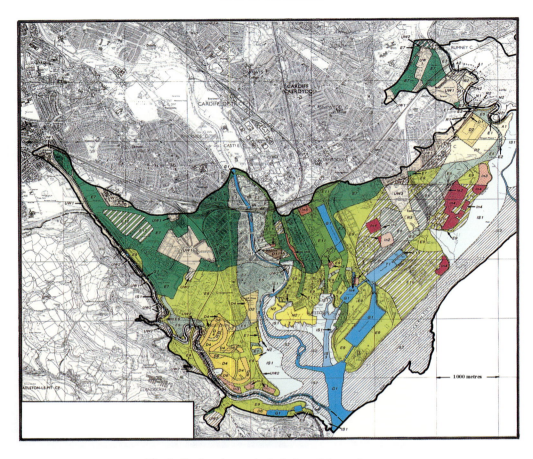

Fig. 3. Engineering geological plan of the study area.

or in press at the time (Allen & Rae 1987; Hawkins 1984; Hornby *et al.* 1993).

Using this primary model format, a search was undertaken of all the available map and borehole record data for the site. A conceptual model of the City of Cardiff was generated that included geological, hydrogeological and anthropogenic data and their functional relationship to the movement and time-variant behaviour of surface and subsurface water bodies. This model was then verified and enhanced by carefully planned investigatory borehole/piezometer work. From this model a robust, logical, qualitative and quantitative risk assessment was derived for peer review (Edwards 1997).

The local ground model

The local model was represented as idealized two-dimensional sections in two forms: first, a model to show the general sequence of strata and their geographic

boundaries (Fig. 1); and second, a model to demonstrate the behaviour of groundwater in three dimensions (Fig. 2).

An engineering geological plan was generated for that area of the city where the main fluvio-glacial sand and gravel aquifer is confined by alluvium (Fig. 3). This illustrated the distribution of strata in three dimensions related to their time sequence deposition as shown on the key to the plan (Fig. 4). Using this information, together with carefully selected ground verification data, it was possible to prepare a functional computer model of the groundwater regime and to use this to examine and verify its current time-variant behaviour. This computerized model was then used to examine the potential and probable impact of the barrage on this regime. This was achieved by applying critical parameter ranges to a series of sensitivity model analyses using rational time periods representing short-term response and steady-state conditions. The groundwater modelling program used was a site-specific modified version of Modflow.

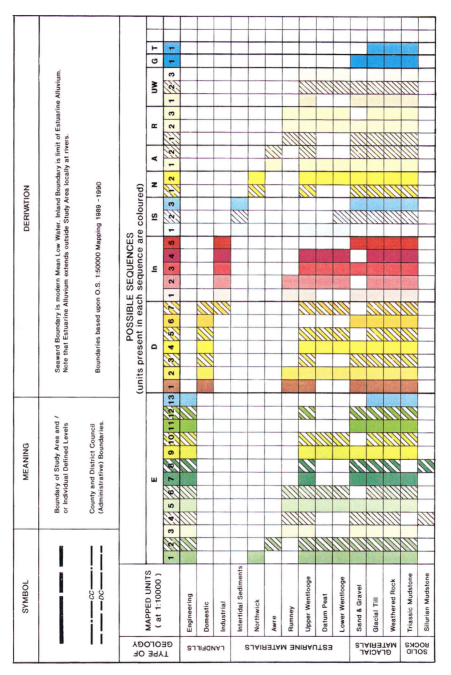

Fig. 4. Legend to accompany the engineering geological plan of the study area (Fig. 3).

Use of ground models

The risk generated by the calculated impact of the barrage was examined using the developed and refined hydrogeological ground model. A rational risk assessment based on a logical scoring system related to nine potential identifiable risk criteria was undertaken. The results were presented as part of the parliamentary proceedings (Tables 1 to 4). The worst case and probable case conditions were assessed. A cost estimate was made for the potential remedial works package required. This risk analysis was developed from the evaluation of these models and was presented for discussion to the various Parliamentary Committee hearings.

Table 1. *Summary of effects on infrastructure*

	Hazard	Conclusion
(i)	Flooding of low ground	Discounted
(ii)	Dampness/wetness or flooding of basements, tunnels and services	Some properties at risk and allowed for in budget costings
(iii)	Solution of minerals or changes in movements of pollutants	
	– corrosion of construction materials	Discounted except for limited risk of sulphate attack, allowed for in budget costings
	– gas displacement	Effects limited to the impounding stage of barrage construction
	– salt leaching	Discounted
	– solution of pollutants	Limited effects on infrastructure, allowed for in budget costings
	– dispersion of pollutants	Discounted but a precautionary programme of monitoring has been recommended
(iv)	Reduction in bearing capacity under foundations and in skin friction of piles	Discounted
(v)	Settlement or swelling of foundation soils	Settlement discounted. Limited effects due to swelling of foundation soils, allowed for in budget costings
(vi)	Structural distress due to increased water pressures on tunnels or basement walls	Discounted
(vii)	Increased difficulty with dewatering excavations during construction	Discounted
(viii)	Surface blowouts if water in aquifer becomes artesian and confining soil is too thin to resist uplift	Discounted
(ix)	Reduction of slope stability	Discounted

Table 2. *Effects of local ground conditions on detrimental effects arising from hazard*

Hazard*	Local ground conditions										
	Fill over			Estuarine alluvium over			Gravel over			Glacial till over	
	Estuarine alluvium	Gravel	Glacial till	Gravel	Glacial till	Rock	Estuarine alluvium	Glacial till	Rock	Gravel	Rock
ii	2	3	1	2	1	1	2	3	3	3	1
iii	3	3	2	2	1	0	0	0	0	0	0
v	2	0	0	2	2	2	2	0	0	0	0

* Hazards: ii, dampness/wetness or flooding of basements, tunnels and services; iii, solution of minerals or changes in movements of pollutants; v, settlement/swelling of foundation soils (settlement discounted as a result of Stage III investigations.
Key: 0, negligible risk of detrimental effects; 1, local conditions will tend to ameliorate detrimental effects; 2, local conditions will tend to accentuate detrimental effects; 3, local conditions will definitely accentuate detrimental effects.
This classification has been developed specifically for conditions encountered in the study area and should not be applied elsewhere. Local conditions addressed in the table encompass all soil profiles encountered in the study area.

Table 3. *Risks with respect to anticipated groundwater change in study area zone*

Zone	Anticipated rise in ambient standing groundwater level (m)	Hazard* ii	iii	v
A	0 to 0.5	0	0	0
B1	0.5 to 1.5	2	1	0
B2	1.5 to 2.5	2	1	1
C	2.5 to 3.0	3	2	2
D	Zero because zone is currently tidal	0	1	0

* Hazards: ii, dampness/wetness or flooding of basements, tunnels and services; iii, solution of minerals or changes in movements of pollutants; v, settlement/swelling of foundation soils (settlement discounted as a result of Stage III investigations).
Key: 0, no risk (but existing problems may persist); 1, some effects possible; 2, some effects probable; 3, perceptible effects will occur in places.
This classification has been developed specifically for conditions encountered in the study area and should not be applied elsewhere.
New equilibrium levels will be below existing maxima in Zone D and in this zone conditions will therefore be improved.

In addition, the original ground and groundwater behaviour models, both of which comprised interrogatable computer databases, were further refined in response to constructive technical criticism by parties opposing the scheme. This was done at the request of Parliament as a precondition of passing the Bill for the project programme.

Both the engineering geological ground model and the computerized groundwater behaviour model are in the public domain and provide a robust, logical and valuable technical archive that can be interrogated for other development projects or purposes. Such models unquestionably offer a simple and effective means of achieving ongoing data refinement. This can be done by updating the databases in the light of additional ground verification. This output can be simply added to the primary computer database.

It is an important principle that by its very nature such information systems should be held in the public domain and should not be 'owned or operated for profit' by any commercial concern or organization. They should also be easily accessible and user friendly in order to maximize their value. It would be unrealistic to anticipate that they should be accessible without charge, but the cost of access should be minimal and involve a *quid pro quo* with respect to data providers in order to encourage participation and model development. The effective ongoing operation and updating of this particular data model presents a management problem that to date remains unresolved.

References

ALLEN, J. R. L. & RAE, J. E. 1987. Late Flandrian Shoreline Oscillations in the Severn Estuary: a geomorphological and stratigraphic reconnaissance. *Philosophical Transactions of the Royal Society. London*, B, **CCCXV**, 185–230.

ANDERSON, J. G. C. & BLUNDELL, C. K. 1965. The sub-drift rock surface and buried valleys of the Cardiff district. *Proceedings of Geologists' Association*, **76**, 367–378.

EDWARDS, R. J. G. 1997. A review of the hydrogeological studies for the Cardiff Bay Barrage. *Quarterly Journal of Engineering Geology*, **30**, 49–61.

HAWKINS, A. B. 1984. Depositional characteristics of estuarine alluvium: Some engineering implications. *Quarterly Journal of Engineering Geology*, **17**, 219–234.

HORNBY, R. P., EDWARDS, R. J. G., RICE, S. M. M., BENTLEY, S. P. & VINING, P. 1993. *The presentation of earth science information for planning development and conservation: illustrated by a study of the Severn Levels*. Department of the Environment, Vol. 1.

Table 4. *Infrastructure types and their sensitivity to potential hazards arising from the proposed barrage*

Hazard*	Light development shallow foundations (low rise) Residential	Industrial	Light development deep foundations/ basements (low rise) Residential	Industrial	Mixed developments (including high rise and basements): Residential/ Industrial	Embankments	Bridges and viaducts	Sports arenas	Buried services at depths >1.5 m	Drains/ sewers at depths >1.5 m	Buried services at depths <1.5 m	Drains/ sewers at depths <1.5 m
ii	Ins	Ins	Det	Det	Det	Ins	Ins	Po	Pb	Det	Ins	Ins
iii	Po	Po	Pb	Pb	Pb	Po	Po	Po	Pb	Det	Ins	Ins
v	Po	Po	Pb	Pb	Pb	Det	Det	Ins	Po	Po	Ins	Ins

* Hazards: ii, dampness/wetness or flooding of basements, tunnels and services; iii, solution of minerals or changes in movements of pollutants; v, settlement/swelling of foundation soils (settlement discounted as a result of Stage III investigations).
Key: Ins, infrastructure is insensitive to hazard; Po, possibly some effects should the hazard arise; Pb, probably some effects should the hazard arise; Det, detrimental effects expected should the hazard arise.
This classification has been developed specifically for conditions encountered in the study area and should not be applied elsewhere.

Pipeline route selection and ground characterization, Algeria

P. G. Fookes[1], E. M. Lee[2] & M. Sweeney[3]

[1] Consultant Engineering Geologist, Winchester, Hampshire, UK
[2] Department of Marine Sciences and Coastal Management, University of Newcastle, Newcastle-upon-Tyne, UK
[3] BP-Amoco Exploration, London, UK

Purpose of the study

A terrain evaluation of a remote area of hyperarid desert in central Algeria was undertaken as part of the front-end engineering design (FEED) studies for a gas field development. The work was part of the identification of suitable route options for a large buried pipeline between the gas fields and the existing pipeline network, some 500 km to the north. The results of the study provided information to support construction costings and preliminary engineering design and construction evaluation of potential geohazards and geotechnical issues, notably trench excavatability and dune mobility, and to help subsequent selection of a specific alignment within the corridor.

Techniques used

The stages involved in the study are summarized in Figure 1, the work formed part of a broader programme of geological studies to characterize ground conditions along a preferred pipeline route. The basis for the preliminary terrain evaluation was a combination of a desk study review and interpretation of a 1:100 000 scale composite (SPOT and Landsat TM) satellite image of the route corridor (approximately 100 km either side of the proposed route). This evaluation was carried out in the UK and was used to identify locations for a limited number of boreholes and trenches. The results from the boreholes and trenches were used, along with a limited 'ground truthing' programme in Algeria, to refine the terrain evaluation and provide typical engineering geological and geomorphological characteristics for each terrain unit.

The terrain evaluation involved the identification of generic *terrain models* (i.e. landscape types) each with a characteristic assemblage of *terrain units*. In many areas terrain units were divided into *subunits*. The terrain units define areas within which certain predictable combinations of surface forms and their associated near-surface materials and geohazards are likely to be found. At the scale of mapping, each terrain unit is assumed to have a consistent range of properties (the average terrain characteristics and their potential variability), i.e. for the purposes of the exercise each unit can be considered as a single unit for route planning or costing.

In most cases the criterion for identifying units and drawing boundaries was a combination of topography and geology. In the study area the landscape could be easily subdivided into a simple sequence of *plateaux* bounded by marked *scarps*, low-lying depressions with almost flat *plains* and *sand seas*. Interpretation of near-surface materials, geology and geomorphological processes was based on experience of landform–process–material relationships in desert environments, plus available geological maps (see below) and information on the geomorphology of the area (particularly duricrusts and sand dunes). Subsequent ground truthing in the field (overflights and fieldwork) has enabled the preliminary evaluation to be confirmed or refined.

In the field drive over many of the terrain characteristics were quickly assessed using a 1–5 numerical scale to describe the relative significance between units and subunits (Table 1). In addition, field assessments were also quickly made of other relevant engineering geomorphological (e.g. wadi dimensions, nature and extent of instability, etc.) and engineering geological (e.g. rock/duricrust strength, fracture spacing, estimated silica content, etc.) features for most terrain units.

Terrain classification

Three main terrain *models* were identified: the Tademait Plateaux (Fig. 2), the Grand Erg Occidental (Fig. 3) and the Hassi R'Mel Plateau. Each model was subdivided into a suite of terrain units and, in most cases, terrain subunits.

On the basis of the satellite image interpretation and the subsequent ground truthing, each terrain unit and subunit was classified by:

1. *Engineering geomorphology, subdivided into:*
   - surface form;
   - geomorphological processes (i.e. water, wind, mass movement, solution weathering, salt weathering, etc.);

From: GRIFFITHS, J. S. (ed.) *Land Surface Evaluation for Engineering Practice*. Geological Society, London, Engineering Geology Special Publications, **18**, 115–121. 0267-9914/01/$15.00 © The Geological Society of London 2001.

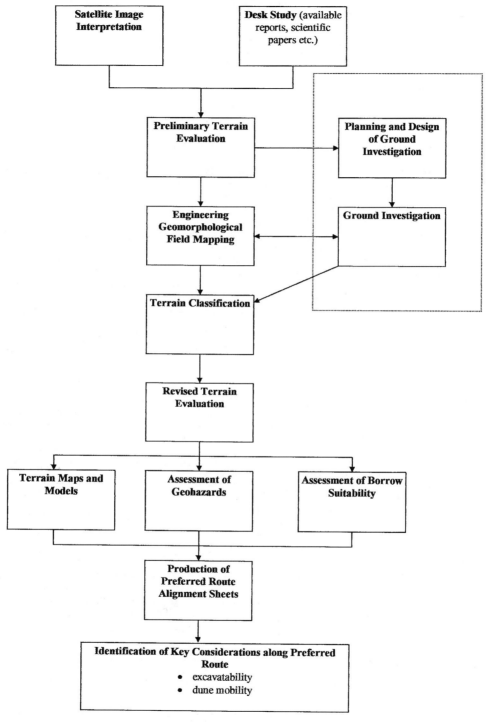

Fig. 1. The work programme.

Table 1. *Terrain characteristic classes*

	Class 1	Class 2	Class 3	Class 4	Class 5
Undulations/relief	Sensibly planar	0–2 m over 0.1–1 km	2–10 m over 0.1–1 km	5–50 m over 0.25–2 km	20–100 m over 0.25–2 km
Fluvial systems	Sheet flow and infiltration	As 1 plus fluvial channels <5 m wide, 1 m deep	As 2 but channels <20 m wide, 3 m deep	As 3 but channels <100 m wide, 5 m deep	As 4 but channels >100 m wide, 5 m deep
Flooding	None	Minor, not significant	Likely, but minor realignment training works required	Very likely, training works/ realignment required	Very likely, best avoided
Trafficability (cf. Landrover)	Reasonable road	Tracks may exist, usually no bogging down. Some need for low ratio 4 × 4	No tracks. All low ratio 4 × 4, some bogging down	As 3 but lots of bogging down and/or many detours required	Not sensibly viable
Slope instability	No instability seen or expected	No or very little instability seen, but in certain circumstances could be expected	As 2 with some instability seen (avoid or take the risk)	Instability commonly seen and expected (avoid or engineer against)	Very extensive chaotic instability. Frequent failure (avoid)
Gypsum heave	None seen or expected	No or very little instability seen, but in certain circumstances could be expected	Heave seen, with vertical movement up to 0.5 m	Heave common, with vertical movement up to 2 m	As 4 but movement >2 m
Pavements and surfaces	Sand sheet surface, no pavement	One layer of gravel, particles not touching	One layer of gravel, particles touching	River alluvium	Fan deposits
Duricrust	None seen or expected	Crust (i.e. BS5930 up to MW, e.g. gypcrete)	Floating crete (i.e. BS5930 >MW, e.g. calcrete, siliceous calcrete)	Caprock (i.e. surface-enriched bedrock)	Crete and caprock

Gypsum heave can also be considered a 'salt aggression' index (1 = no problem; 5 = severe problem), but salt aggression also depends on moisture and other salts (e.g. chlorides).

- hazards (e.g. flooding, instability and subsidence, dune mobility, aggressive soils, etc.);
- superficial materials;
- duricrusts;
- bedrock.

2. *Engineering geology, subdivided into*:
   - rock mass structure;
   - fracture spacing;
   - silica class;
   - rock strength;
   - potential excavatability;
   - suitability for borrow materials;
   - trafficability.

3. *Identified principal constraints to route planning.*

A number of simple models were also developed to demonstrate the relationships between topography, terrain units and particular terrain characteristics: flood hazard, duricrusts, slope instability and gypsum mobility and heave (e.g. Fig. 4a and b).

Main findings

The principal geological and geomorphological factors that influenced route selection and costing were:

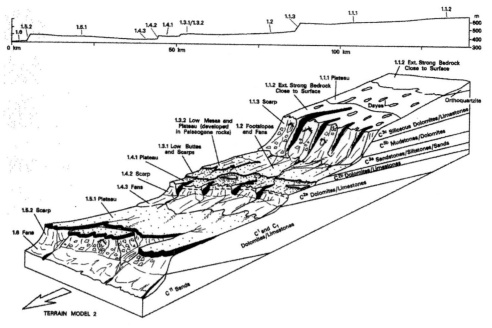

Fig. 2. Terrain model 1: Tademait plateaux.

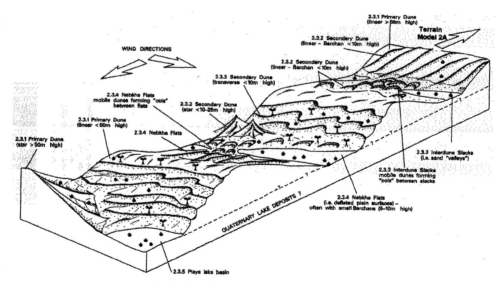

Fig. 3. Terrain model 2. Grand Erg Occidental.

1. the presence of very strong to extremely strong, massive bedrocks and duricrusts with high silica content in parts of the area. As a result there is potential for very slow trenching rates (blasting may be required in some sections) and high pick (tooth) consumption for rock cutting trenching machines;

2. the presence of mobile dunes, generally 5–15 m high, within the sand sea. From world-wide dune migration research (e.g. Cooke *et al.* 1992), it is expected that these dunes may move by around 10–15 m/year;

3. the presence of widespread instability on scarp slopes;

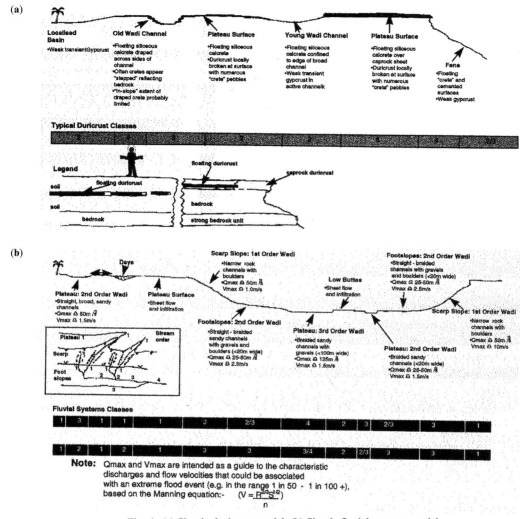

Fig. 4. (a) Simple duricrust model. (b) Simple fluvial systems model.

4. the potential for localized scour and exposure of the pipeline in wadi channels, during rare flood events;
5. the potential for subsidence or collapse over the circular, silty sand filled depressions on many plateaux surfaces (dayas);
6. the potential for gypsum heave and aggressive ground conditions;
7. the widespread availability of potentially suitable materials for trench backfill after processing (e.g. stone pavements, fan deposits, etc.), road construction (e.g. gypsum-rich desert soils) and concretes.

The most important factors, trench excavatability and dune mobility, are considered below.

Excavatability

Trench excavatability is controlled by the nature of the rock/duricrust (notably strength and fracture spacing) and the equipment used (size, type and method of working). Two geotechnical-related issues are central to estimating costs of trenching operations, namely the rate of progress and the wear of the cutting tools (bits/picks/teeth). Characteristic rock/duricrust strength and fracture spacing measurements were estimated for each terrain unit and subunit as part of the fieldwork programme. This information was supplemented by point load test data collected during the parallel borehole investigation. For each terrain unit, this information

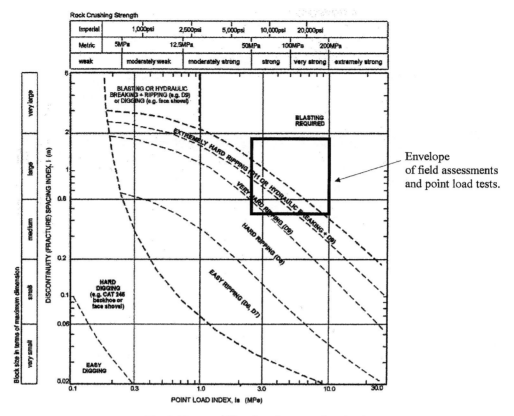

Fig. 5. Excavatability chart for a terrain subunit.

was plotted on an excavatability chart (developed from Pettifer & Fookes 1994) to determine the relative progress and potential mode of operation for particular machines. An example of this is presented as Figure 5.

An estimate was made of the relative pick (i.e. bit/tooth) consumption in different terrain units, based on the relative hardness of the rocks/duricrusts and the picks (this estimate is specifically for trenching machines although it is broadly applicable for conventional ripping operations as an indication of relative tine wear). Rocks/duricrusts comprising quartz (Vickers hardness $= 11\,000$ MPa), or harder minerals, will cause significant abrasion of the picks (assumed to be tungsten carbide, Vickers hardness $= 10\,000$–$18\,000$ MPa). Thus, trenching in those areas with silica-rich materials (i.e. quartz-rich) will cause the highest relative rates of pick consumption.

Dune mobility

The Grand Erg Occidental is a complex sand sea, covering some $100\,000\,\text{km}^2$ (see McKee 1979; Callot

1988) formed during periods of extreme aridity and high wind speeds in the Pleistocene. Along the proposed pipeline route, a range of dune features can be observed at different spatial and temporal scales.

1. Ephemeral dunes (small, short-term features): probably develop and decay in response to the annual wind climate or even daily wind climate. Predominantly transverse dunes. Actively mobile. Probably <5m high.
2. Secondary dunes (larger, medium-term features): probably develop in response to 10–100 year variations in the wind climate. They can probably withstand minor seasonal variability and are shaped by extreme events. Predominantly transverse, barchan and star dunes. Probably medium sand. Episodically active. Generally 5–15 m high.
3. Primary dunes (largest, long-term features): reflect adjustments over very long timescales (e.g. 1000–10 000 years). Essentially relict linear features from the arid phases in the Pleistocene. These features generally have sparsely vegetated plinths (low angled

side-slopes) and are probably of coarser material than the ephemeral and secondary dunes. They are immobile under current climatic conditions and probably up to 100 m high.

Amongst the issues relevant to the consideration of route options through the sand sea are:

- the potential for pipe failure due to loading by blown sand;
- the potential for exposure of the pipe through wind and/or water scour, leaving lengths of pipe unsupported;
- the identification of efficient routes through the dune field to minimize cut-and-fill operations.

The potential hazards associated with rapid dune movement were considered, using a series of empirical sand transport models. A helicopter-based reconnaissance of ground conditions was made within the sand sea to determine whether it was possible to define potential routes which maximize the use of interdune flats and slacks, with crossing of active dunes limited to around 10–30% of selected alignment.

Lessons learnt

The objective of the study was to produce an overview of the terrain conditions along the pipeline corridor, concentrating on potential geohazards and trench excavatability. The work was mainly based on satellite image interpretation and less than 20 man-days field survey. Large lengths of the proposed route were only observed from vehicle or from an aircraft. The techniques involved are straightforward. However, the successful completion of the exercise within the significant logistical constraints imposed by the security problems in Algeria and the very tight timescale (routing information was required within three months of the start of the studies) relied heavily on the judgement and experience of the authors in satellite image interpretation and terrain evaluation in desert environments.

When considering the value of the terrain evaluation approach it is useful to compare it with other, perhaps more traditional, approaches to pipeline route selection. The two approaches considered for delivering information on this pipeline route were:

1. Approach A: a systematic borehole investigation, with boreholes planned, on average, every 1 km, supported by a walk-over survey;
2. Approach B: a combination of terrain evaluation methods and a limited number of boreholes and trenches (around 50 in total) specifically located to test the terrain models.

The terrain evaluation approach (Approach B) was considered to be the only method that could deliver the necessary information within the required timescale. It is estimated that it has provided information that is 'fit for purpose' (i.e. supporting 'front-end' cost estimates) at around one-tenth of the cost of a systematic borehole investigation (Approach A). The terrain evaluation approach also provides a structural view of the principal cost drivers associated with pipeline construction through the area and, hence, a framework for considering alternative route options, whereas the information provided by a borehole investigation would be specific to a chosen alignment.

References

CALLOT, Y. 1988. Evolution polyphasée d'un massif dunaire subtropical: Le Grand Erg Occidental (Algérie). *Bulletin de la Société géologique de France*, **4**(4), 1073–1079.

COOKE, R. U., WARREN, A. & GOUDIE, A. S. 1992. *Desert Geomorphology*. University College Press, London.

MCKEE, E. D. (ed.) 1979. *A Study of Global Sand Seas*. USGS Professional Paper **1052**.

PETTIFER, G. S. & FOOKES, P. G. 1994. A revision of the graphical method for assessing the excavatability of rock. *Quarterly Journal of Engineering Geology*, **27**, 145–164.

Building the geological model: case study of a rock tunnel in SW England

P. G. Fookes[1] & D. T. Shilston[2]

[1] Consultant Engineering Geologist, Winchester, Hampshire, UK
[2] W. S. Atkins Consultants Ltd, Epsom, Surrey, UK

Objectives

A 'geological model' is a representation of the geology of a particular location. '*The form of the model can vary widely and include written descriptions, two-dimensional sections or plans, block diagrams, or be slanted towards some particular aspect such as groundwater or geomorphological processes, rock structures and so on*' (Fookes 1997, p. 294). Formal creation of a geological model is one of the fundamental processes by which geologists, geomorphologists and other Earth scientists assemble an understanding of the ground conditions at a site. It is a powerful and cost-effective vehicle for conveying this understanding, often in simplified form, to other disciplines such as civil and structural engineers and planners.

Geological models are not always easy or straightforward to create. This is particularly so at the desk study and field reconnaissance phases of site investigation. However, it is during these early phases that a model (or models) can be particularly useful by helping to set out what is known, what is conjectured, and where significant gaps in knowledge may lie. Geological fieldwork provides important information for the model, yet much of the geological interpretation of such fieldwork is necessarily subjective. The case study described here illustrates how the vagaries of geological exposure and ground investigation programmes can be evaluated to give an understanding of the completeness and reliability of such data. This evaluation of data is called here the 'determinability' of the geology.

The project

This case study concerns a large shallow road tunnel in southwest England, which was subject to arbitration that was eventually settled out of court. The tunnel was constructed through a variably weathered rockmass comprising steeply dipping slates which contained layers and irregular bodies of igneous rock.

Techniques

Four types of geological field data were available prior to construction of the tunnel (Fig. 1):

- coastal exposures;
- railway cuttings;

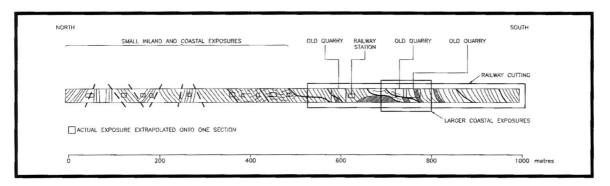

Fig. 1. Geological section determined from exposures near line of tunnel.

From: GRIFFITHS, J. S. (ed.) *Land Surface Evaluation for Engineering Practice*. Geological Society, London, Engineering Geology Special Publications, **18**, 123–128. 0267-9914/01/$15.00 © The Geological Society of London 2001.

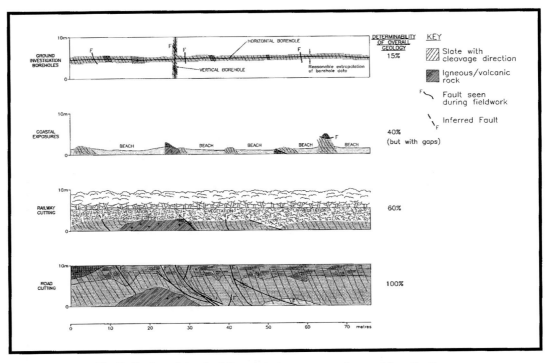

Fig. 2. Geology determined from actual exposures and ground investigations.

Table 1. *Determinability of geology from different sources of information*

Location	Approx. age (years)	Estimated physical visibility of rock (%)	Ability to determine large-scale geology (geological domains) estimate (%)	Ability to determine small-scale geology (geological styles) estimate (%)	Total out of 300%	Overall percentage of determinable geology (%)	When determinable?
Ground investigation							
Boreholes	–	10	30	5	45	15	Reasonable at time of tender
Trial adit	–	75	75	90	240	80	Reasonable at time of tender
Surface exposures							
Coastal exposures	2000	10	80	30	120	40	Reasonable at time of tender
Railway cutting	150	25	100	50	175	60	Unreasonable at time of tender
Road cutting	9	100	100	100	300	100	Not available at time of tender
Relevance to anticipated construction			Determines overall methodology	Determines daily performance			

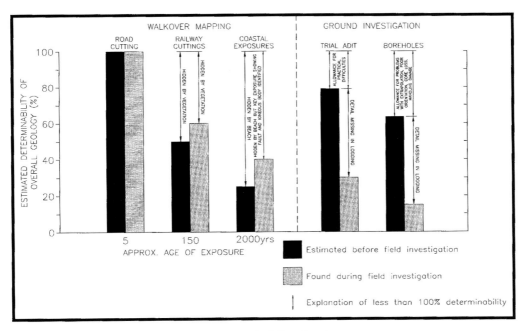

Fig. 3. Determinability of geology.

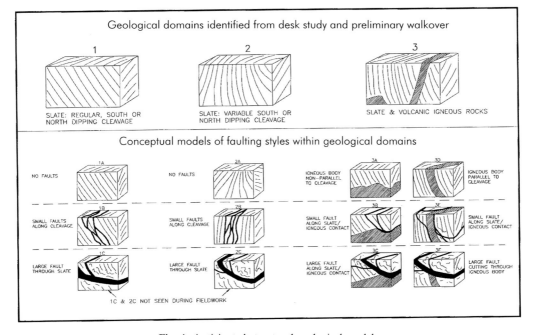

Fig. 4. Anticipated structural geological models.

- abandoned quarries;
- pre-construction boreholes and a trial adit for the road tunnel.

A fifth set of field data was available during and following construction (Fig. 1):

- road cutting (leading to one of the tunnel portals), about 10 years old.

For contractual reasons, the work described here was carried out following construction. Field data were gathered by the authors from geological exposures using conventional engineering geological mapping techniques appropriate for geological reconnaissance. The information was placed in a broader context using the maps and memoir of the British Geological Survey (Ussher 1907) and later publications (e.g. Coward & McClay 1983). Logs of the pre-construction trial adit and boreholes drilled for the tunnel project were reviewed, but the drill core was not relogged.

Each type of exposure, the adit and the drill core had different characteristics that can be thought of as different samples of the geology of the area. It became apparent during the work that, because of these characteristics (such as age and visibility of the rock mass), each would present different understandings (or models) of the geology.

Results

Bedrock in the area of the tunnel consists of metamorphosed Upper Devonian marine sediments (predominantly slates) with intercalated igneous rocks (tuffs, dykes and lavas) which were folded and faulted mainly during the Hercynian Orogeny. Ussher (1907) shows thrust faults dipping to the south and roughly coaxial with the cleavage of the slates. More recent research has highlighted the importance of thrust tectonics in the Hercynian deformation of the area (Coward & McClay 1983) and, whilst no local major thrust faults are marked on the current 1:50 000 British Geological Survey map, small-scale thrust faults are likely to be present. Also in the area of the tunnel, Dearman (1963) recognized a zone of wrench faulting trending north-northwest to south-southeast, which could be extrapolated to the tunnel location. However, evidence for this extrapolation is sparse. Rocks younger than the Devonian do not outcrop in the area adjacent to the tunnel. The area was subjected to very long periods of subtropical weathering during the Tertiary, followed by the periods of periglacial activity during the Quaternary; these conditions are particularly relevant as the tunnel was at shallow depth.

Incompleteness of geological information at various exposures may not always be recognized and searching for exposures is often time-consuming and frequently

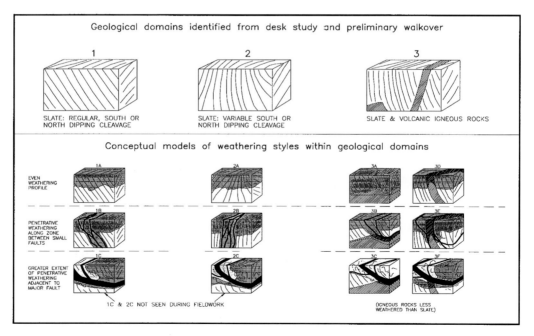

Fig. 5. Anticipated weathering models.

unrewarding. However, in this case not only were more exposures found than were anticipated by inspection of topographic and old geological maps, but also there was a relationship between their condition (determined largely by the age of the exposure) and the information which they yielded.

Figure 1 is a simple section of the location and extent of exposure in the area around the tunnel and the geology that was interpolated between the locations. Care was taken to ensure that interpolations were reasonable and took account of the distances between exposures and their location with respect to the structural grain of the bedrock. Figure 2 shows the geology that could be determined from four specific types of exposure:

- unorientated vertical boreholes at about 50 m spacing and horizontal boreholes connecting the two portals (from the tunnel site investigation proper);
- coastal exposures from behind a shingle beach, crudely estimated to be around 2000 years old;
- a railway cutting about 150 years old, now partly vegetated and covered with slope wash debris;
- a new road works cutting in good condition.

Figure 2 also shows estimates of the geological information gleaned by the geological fieldwork. Table 1 summarizes these estimates and illustrates a simple concept of the 'determinability' of the local geological conditions at the four different types of exposure. It shows how the influence of weathering and erosion by the current climate of southwest England clearly reduces the ability to collect data from natural and man-made exposures. It also illustrates the difficulties in obtaining a complete picture from boreholes (especially vertical boreholes) in the cleaved and very steeply dipping Devonian age rocks, particularly information on discontinuities which

was of critical importance for the tunnel's design and construction. Figure 3 shows some of this information portrayed as a histogram to illustrate how the ability to make observations on discontinuities decreases with increasing age of exposure.

Figures 4 and 5 show various systematic attempts at building three-dimensional (block diagram) geological models of potential situations with all the information available both from boreholes and from the mapping fieldwork. Figure 5 is particularly important for a shallow tunnel. It illustrates the effects of differential penetrative weathering along the very steeply dipping cleavage, faults and igneous bodies. The problems of investigating such rockmasses using boreholes without having a clear geological model or models in mind can be readily appreciated.

Figure 6 shows an attempt to integrate into a single model the various items of geological information available prior to construction of the road tunnel, excluding the boreholes and trial adit. It is reproduced from the Masters thesis of a student who studied this case history after the conclusion of the arbitration proceedings (Anderson 1996). The student carried out his own independent desk study and reconnaissance fieldwork. His model illustrates the simple way in which geological conditions can be portrayed in two dimensions, and in which geological hazards and gaps in knowledge can be highlighted, leading to improved planning of ground investigation work (boreholes, etc.).

In summary, our conclusions were as follows.

- The tunnel was shallow, with a complex interplay of differential weathering and changes in attitude of cleavage; as an additional complexity, it

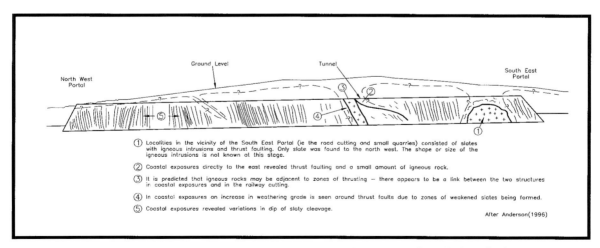

① Localities in the vicinity of the South East Portal (ie the road cutting and small quarries) consisted of slates with igneous intrusions and thrust faulting. Only slate was found to the north west. The shape or size of the igneous intrusions is not known at this stage.

② Coastal exposures directly to the east revealed thrust faulting and a small amount of igneous rock.

③ It is predicted that igneous rocks may be adjacent to zones of thrusting — there appears to be a link between the two structures in coastal exposures and in the railway cutting.

④ In coastal exposures an increase in weathering grade is seen around thrust faults due to zones of weakened slates being formed.

⑤ Coastal exposures revealed variations in dip of slaty cleavage.

After Anderson(1996)

Fig. 6. Model of geology along tunnel route as anticipated by Anderson (1996) from site investigation information.

had faults which were difficult to predict and thin igneous bodies which appeared effectively impossible to predict.

- It was apparent from the model-building exercises that more borehole information would have been necessary to portray a sufficiently accurate picture of the eventual tunnel conditions.
- Without the development of such geological models, and an understanding of the determinability of the geology, the borehole information would have been considered adequate.
- The modelled conditions strongly suggested that despite otherwise reasonable borehole coverage at about 50 m centres, together with long horizontal boreholes to improve the geological knowledge, more boreholes and/or a contract with reference conditions and financial provision for the observational method would have been necessary.

Acknowledgements. The authors thank Balfour Beaty Civil Engineering Ltd for permission to publish work carried out on their behalf. We would also like to thank Mr Andy Lewis, engineering geologist, who worked on the original arbitration case, and Mr Darren Anderson, whose MSc thesis re-evaluated some of the original data.

References

ANDERSON, D. 1996. *The Application of Geology in Optimising Site Investigation Layout Techniques*. MSc Thesis, Queen Mary and Westfield College, University of London.

COWARD, M. P. & McCLAY, K. R. 1983. Thrust tectonics in S. Devon. *Journal of the Geological Society*, London, **140**, 215–228.

DEARMAN, W. R. 1963. Wrench faulting in Cornwall and South Devon. *Proceedings of the Geologists' Association*, **74**, 265–287.

FOOKES, P. G. 1997. The First Glossop Lecture: Geology for Engineers: the Geological Model, Prediction and Performance. *Quarterly Journal of Engineering Geology*, **30**, 293–424.

USSHER, W. A. E. 1907. *The Geology of the County Around Plymouth and Liskeard*. Memoirs of the Geological Survey, HMSO, London.

Development of a ground model for the UK Channel Tunnel portal

J. S. Griffiths

Department of Geological Sciences, University of Plymouth, Devon, UK

Purpose of survey

Engineering geomorphological mapping is primarily concerned with identifying and mapping the features on the ground surface. However, detailed mapping linked to limited subsurface data can provide the first approximation of a ground model in some situations. In this example mapping was carried out at the location of the proposal Channel Tunnel portal at Castle Hill near Folkestone, England, prior to any detailed ground investigations with the aim of establishing the nature and extent of the landslide complex that existed at the site. Full details of the study can be found in Griffiths *et al.* 1995.

The site

The Channel Tunnel terminal and portal on the UK side is located immediately below the Etchinghill escarpment, a scarp slope developed in the Lower Chalk of the North Downs to the west of Folkestone. Whilst the tunnel itself was positioned within the lower permeability zone of the Chalk Marl, the works for the terminal and portal required excavations in the Lower Chalk, the underlying Gault Clay and through the solifluction cover of Coombe Rock.

The 1:10 560 scale BGS map identified six landslide complexes within the terminal and portal works area (Aarons *et al.* 1977), although there was no information available on the form, depth and current stability of these features. The design for the main tunnel portal, however, required it to enter the hillside through the centre of one of the landslides at Castle Hill.

Techniques used

Field mapping of the geomorphology was undertaken at a scale of 1:500 using standard procedures for geomorphological survey work (Brunsden *et al.* 1975). The engineers were able to provide large-scale plans with contours at 1 m intervals as a base. These plans made detailed mapping much easier and almost certainly increased the accuracy of the geomorphological boundaries shown on the final maps.

The conceptual model

Castle Hill Landslide had a highly complex morphology (Fig. 1). The large-scale morphological mapping was followed by a geomorphological interpretation that utilized the results of a preliminary site investigation to provide the basis for a subdivision of the landslide into five zones.

1. Main landslide backscar with slopes in the range from 35° to 40°. The extent of the backscar was somewhat confused by a Norman defensive ditch which appeared to have been utilized and cleaned out during the Second World War.
2. Benches with minor front scarps within the landslide complex representing displaced blocks associated with the main landslide movements. Five distinct landslide blocks were identified reaching a maximum width of 20 m with their upslope extent clearly indicated by a sharp concave break of slope.
3. Three secondary landslides were identified within the landslide complex with distinctive backscars and downslope accumulation areas, the most apparent being a shallow translational debris slide formed on the southeastern part of the main landslide.
4. The landslide accumulation zone to the west of the Castle Hill Road. Two separate terraces/benches were visible with slopes of 3–6° separated by a 40 m wide scarp with a slope of 7–10°. A 20 m wide scarp that had slopes of between 8 and 13° also marked the front of the lower terrace/bench. The base of this lower scarp was taken as the toe of the landslide on the BGS 1:10 560 scale maps of the area.
5. Made ground. The study of historical maps had shown that there was a significant amount of made ground within the landslide complex.

This geomorphological interpretation of the morphological form of the Castle Hill landslide is presented in Figure 2.

What the survey established

Identification of the various components of the landslide unit provided a basis for examining both its current level of stability, the timing of the original and any subsequent landslide movements and the likely form of

From: GRIFFITHS, J. S. (ed.) *Land Surface Evaluation for Engineering Practice*. Geological Society, London, Engineering Geology Special Publications, **18**, 129–133. 0267-9914/01/$15.00 © The Geological Society of London 2001.

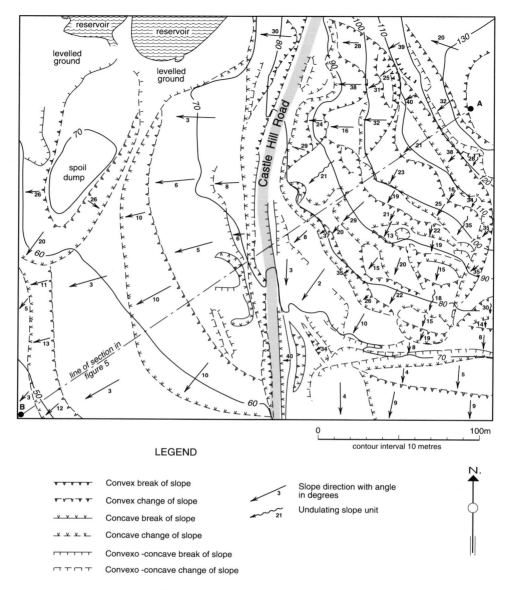

LEGEND

▼▼▼▼▼	Convex break of slope
⌐▼⌐▼▼	Convex change of slope
ᴠ ᴠ ᴠ ᴠ	Concave break of slope
⌐ᴠ ᴠ ᴠᴸ	Concave change of slope
⌐⌐⌐⌐⌐	Convexo -concave break of slope
⊓ ⊤ ⊓ ⊤	Convexo -concave change of slope

Slope direction with angle in degrees

Undulating slope unit

N.

Fig. 1. Morphology of Castle Hill landslide.

basal and any intermediate shear surfaces. In addition, following the detailed geomorphological mapping of the site, existing borehole data were compiled and some trial pits in the accumulation area were excavated. The intention was to provide a provisional interpretation of the mode of failure of the main landslide prior to the commencement of the main ground investigations (see cross-section, Fig. 3).

The boreholes along the line of section contained multiple shear surfaces, shear zones and disturbed lengths of Chalk Marl and Gault Clay. This clearly established the landslide as having multiple units and also that there had been more than one factor in the development of shear surfaces. It was concluded that four separate types of shear surface existed within the landslide complex.

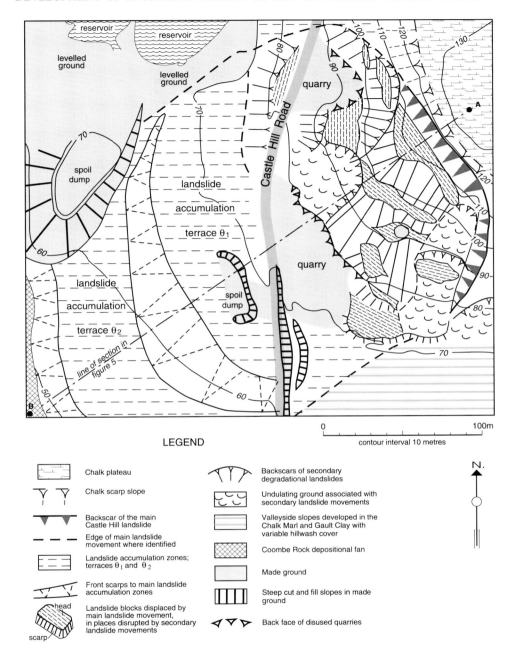

Fig. 2. Geomorphological interpretation of Castle Hill landslide.

1. Type 1: deep-seated shear surfaces in the Gault Clay associated with both the tectonic history of the deposit and the effects of erosion of the overlying Chalk producing unloading shears which resulted in lateral expansion and extrusion of the clay.

2. Type 2: the main surfaces associated with the original Castle Hill landslide. This appeared to have failed in the form of a multiple rotational unit followed by subsequent block disruption, of the form described by Brunsden & Jones (1976) on

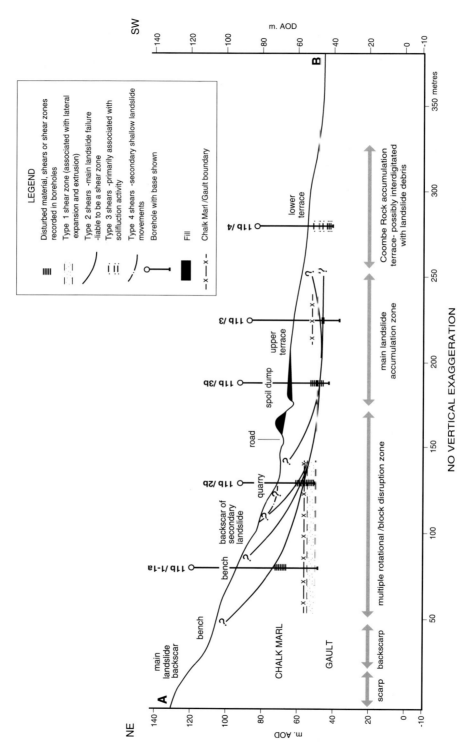

Fig. 3. Cross-section of the Castle Hill landslide.

Stonebarrow in Dorset. The basal shear surface was estimated to be approximately 20 to 25 m below the present ground surface.

3. Type 3: shears in Coombe Rock gelifluction deposits located towards the toe of the main landslide. The relationship between the downslope limits of the land-slide and Coombe Rock deposition was a complex one. Subsequent logging of exposures during construction indicated that in places the Castle Hill landslide had been thrust over the Coombe Rock but then was buried by subsequent Coombe Rock deposits. This gave clear evidence of both the antiquity of the landslide movements and their repeated movement during the post-glacial period.

4. Type 4: the shear surfaces associated with the secondary landslides. These would be continuous but much shallower than the basal shear surface associated with the main landslide movement and represent part of a general degradation process.

This information and the postulated form of the landslide is presented in section in Figure 3. A trial shaft and adit was excavated into the landslide and this confirmed the multiple rotational form with the basal shear surface located at a maximum depth of 24 m below the present ground surface.

The actual distal limits of the landslide were not finally confirmed until construction excavations took place. These established that the scarp downslope of the lower terrace (Fig. 2) was the toe of the landslide.

Similar applications

Geomorphological mapping for civil engineering is not a new concept. However, the extent of the work carried out for the Channel Tunnel was far greater than just a mapping exercise. Historical studies prior to the mapping provided a vast amount of information about man-made features in the landscape which, at the time of investigation, were highly degraded. The extent and detail of the geomorphological mapping also allowed subsurface data to be tied into specific boundaries in geomorphological units in a way that has rarely been possible elsewhere.

For the Castle Hill landslide, because of the critical nature of the engineering works, a highly detailed three-dimensional model of the subsurface geometry was needed. Whilst subsequent detailed, and very expensive, ground investigations refined the understanding of the landslide complex, the model developed from the author's original mapping and interpretation proved to be extremely robust and formed the basis for focusing the subsequent geotechnical investigations into the critical areas.

This type of assessment has many applications in detailed survey work for engineering. Any form of instability which might affect engineering works could be subject to detailed mapping in order to provide both a framework for subsequent detailed ground investigations and a preliminary ground model suitable for initial design assessments.

Acknowledgements. The author would like to TML for permission to publish the work on the Channel Tunnel, and John Abraham from the University of Plymouth for producing the figures in this paper.

References

AARONS, A., WEEKS, A. G. & PARKES, R. D. 1977. Site investigation for the Channel Tunnel British Ferry Terminal. *Ground Engineering*, May, 43–47.

BRUNSDEN, D. & JONES, D. K. C. 1976. The evolution of landslide slopes in Dorset. *Philosophical Transactions of The Royal Society of London, Series A*, **283**, 605–631.

BRUNSDEN, D., DOORNKAMP, J. C. FOOKES, P. G., JONES, D. K. C. & KELLY, J. M. H. 1975. Large scale geomorphological mapping for highway engineering. *Quarterly Journal of Engineering Geology*, **8**, 227–53.

GRIFFITHS, J. S., BRUNSDEN, D., JONES, D. K. C. & LEE, E. M. 1995. Geomorphological Investigations of the Channel Tunnel Terminal and Portal. *The Geographical Journal*, **161**, 275–284

Low-cost road construction and rehabilitation in unstable mountain areas

G. J. Hearn

Scott Wilson Kirkpatrick & Co Ltd, Basingstoke, Hampshire, UK

Rationale

When designing the construction or rehabilitation of roads in unstable mountain regions subjected to floods, landslides, erosion and earthworks failures, it has become usual practice to employ a selection of the techniques described in Section 2. The need for adequate land surface evaluation in this context is obvious: (i) the information so produced is essential to many if not most design processes; and (ii) proper evaluation of topography, materials and geohazard is central to the performance of the road construction, and its maintenance in the longer term. These concepts are embodied in Fookes *et al.* (1985) and in Overseas Road Note 16 (Transport Research Laboratory 1997) which deals specifically with geohazards and road design in unstable mountain areas.

The selection of the techniques, and the manner in which they are applied, will depend on project area conditions, the availability of the necessary data or documents, and the nature of the engineering scheme being proposed. The techniques are usually applied in a progressive manner, with the need to refine and detail ground conditions more closely as the site selection and design procedure takes place. The conclusion to this process is the usual inevitability that final design will not be established until ground conditions are fully defined during construction.

In the case of road rehabilitation projects, existing cut slope exposures provide a more or less continuous record of soil and rock conditions above the road, while the stability of slopes and the observed distress to the existing road pavement provides a useful overview of stability, and the reaction of slopes and drainage channels to road works in general. These benefits obviously do not accrue in the case of new road construction.

The design case studies for a proposed new road in Nepal and the rehabilitation of an existing road in the Philippines are described below to illustrate the value of land surface evaluation techniques in each case.

Construction design of the Arun Access Road, Nepal

The area model

The Arun Access Road was designed to provide access through the remote interior of the Middle Himalaya to the site of a proposed hydropower installation. The Project Area (here defined as the area containing all plausible alignment alternatives) is approximately 5000 km^2, and is located entirely in the drainage basin of the Arun River in east Nepal; this major river is one of five that have cut through the Himalayas, thus maintaining their southerly course as the mountain ranges have risen. To illustrate the extreme difficulty that the topography poses to road alignment, one of the two main alignment corridors considered was forced to make a cumulative rise and fall of 8800 m, equivalent to the height of Mount Everest above sea level, over an alignment distance of 223 km. Slopes are usually steep and underlain by weathered and intensely fractured rock. Erosion and slope instability, predictably, constitute significant hazards.

Desk study identification of alignment corridors

Published 1:25 000 and 1:50 000 scale topographical maps, in conjunction with the interpretation of existing 1:20 000 scale black-and-white aerial photographs, were used to define broad route corridors. This was achieved by identifying acceptable topography from an alignment design point of view, and by locating favourable facets or features in the landscape, such as high level river terraces, potential river crossings, preferred locations for crossing ridges, and slopes suitable for climbing sections (stacked hairpins). Unfavourable features, such as cliffs, complex and deeply incised topography, landslides, areas of slope erosion, eroding river terraces and flood-prone areas, were also identified on the aerial photographs. Because it is rare to find an alignment corridor which is devoid of these features, a careful assessment of risk potential is required through ground reconnaissance.

From: GRIFFITHS, J. S. (ed.) *Land Surface Evaluation for Engineering Practice*. Geological Society, London, Engineering Geology Special Publications, **18**, 135–141. 0267-9914/01/$15.00 © The Geological Society of London 2001.

Ground reconnaissance of potential route corridors

Each of the potential route corridors identified from the desk study was examined in the field. Soil and rock types, slope angles, landslides, colluvial deposits, cliffs, river terraces, etc. were mapped and plotted onto published 1:25 000 scale contour maps. The alignment corridor options were plotted onto the contour maps and onto enlarged paper prints of the aerial photographs, and verified in the field using hand-held levelling equipment. Field notes describing geology, soils, drainage, geomorphology and geohazard were recorded on the contour maps and later summarized according to the land surface classifications shown in Figure 1.

This combination of intensive desk study and field mapping techniques provided maximum familiarity with the terrain at the earliest possible stage in the route corridor selection process. The approach did not rely on landslide hazard mapping in a formal sense, as this technique is not generally compatible with the need for rapid and reliable engineering assessment of route practicability, stability and relative cost.

The collected data were converted and tabulated in the office. Each alignment corridor was summarized

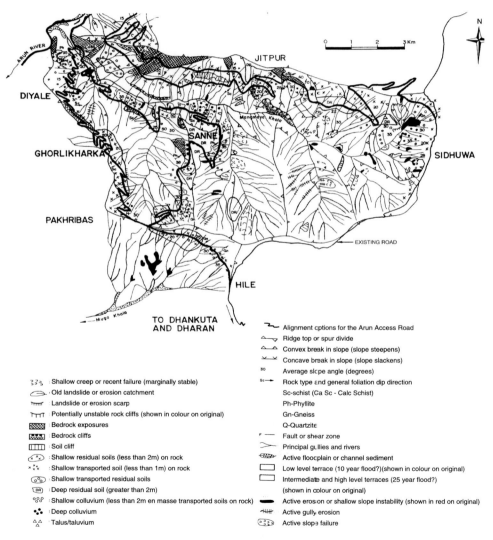

Fig. 1. Comparison of route corridor options, originally at 1:25 000 scale.

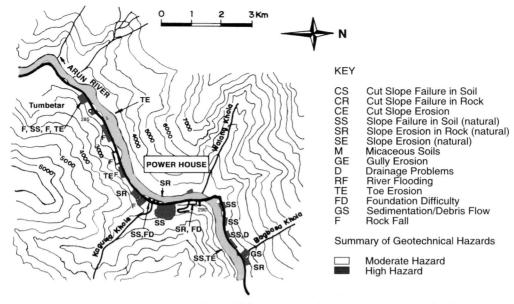

Fig. 2. Summary map showing potential slope and drainage hazards at a scale of 1:75 000 for client familiarization.

according to estimated length, side slope angles, estimated proportions of soil and rock in excavations, and the length of alignment encountering existing and potential slope instability and erosion (potential instability was assessed in a qualitative way by reference to recorded rock mass characteristics, soil types, drainage and slope angle). Broad cost comparisons were made between different alignments using estimated quantities of rock and soil cut, numbers of culverts and lengths of bridge spans, anticipated lengths and heights of retaining wall based on side slope angle and preferred cross-section, and provisional sums for erosion protection and slope stabilization. Summary maps were prepared for the client showing the extent of anticipated slope and drainage hazards along the alignment options at a scale of approximately 1:75 000 (Fig. 2).

Alignment design

Colour and black-and-white aerial photographs were taken along the chosen alignment corridor in association with the Transport Research Laboratory, UK. The photographs were produced at a scale of approximately 1:5000 and covered an area generally 2 km either side of the proposed alignment. The horizontal and vertical alignments were optimized using a photogrammetrically derived ground model corrected for distortion by ground control. The suitability of the alignment was checked using aerial photograph interpretation. The resultant centre-line was then set out on the ground. Plan and vertical profile drawings were prepared with cross-sections printed at 20 m intervals.

Field assessment

A highway engineer and an engineering geomorphologist undertook a systematic assessment of the designed alignment. Shade, tree cover and the inevitable scale distortion on photographs taken in such steep terrain led to contour inaccuracies in some areas. In one location, for instance, a ravine approximately 50 m deep and 50 m wide was unseen on the aerial photographs. It took almost a week of survey through dense jungle to 'find' a suitable alignment that avoided this feature. Less radical decisions to shift the centre-line to the right or to the left were based on the need to avoid unstable or eroding slopes, or to negotiate difficult topography.

Field assessment of ground conditions often led to the conclusion that it would be preferable to construct the road either wholly in cut or wholly on retaining wall, depending upon the anticipated strength, stability and volume of excavated materials, or the suitability of the topography and foundations below the road for wall construction. The centre-line was shifted accordingly, but usually within 20 m of the designed alignment, and frequently less. Engineering geomorphological plans were prepared as part of this exercise and aerial photographs were used in the field to assist in geomorphological interpretation, especially in areas of minor realignment.

Output review

Engineering geomorphological mapping, even at the time of detailed alignment review, was therefore undertaken mostly as a means of recording features and ground conditions. This work led to a decision to realign the centre-line and to provide information on landform, stability of slopes and drainage channels, preferred cross-section and the general suitability of materials for construction. In most cases, the degree of geomorphological detail portrayed on the maps was limited to the delineation of specific features only; there was considered to be little point in carrying out detailed morphological or geomorphological mapping once a design decision had been made and the requirements of the geomorphological assessment had been satisfied. Comments were made on the need for erosion protection, slope drainage and slope stabilization works. These observations were included in the design report and ancillary documentation, and the recommendations for remedial works were incorporated as lump sum items in the Bill of Quantities.

Rehabilitation design of the Halsema Highway, Philippines

The area model

The Halsema Highway follows the spine of the steeply sloping Central Cordillera in north Luzon. The topography is composed of intrusive and extrusive volcanic rocks, mostly diorites, basalts and breccias, that are highly fractured and altered, together with limestones, shales and conglomerates. Three main active fault structures cross the alignment and form part of the Philippine Fault which defines one of the most active plate margins in the world. In July 1990 an earthquake which registered 7.8 on the Richter Scale caused major damage to the road and its adjacent slopes, as well as structural damage and loss of life in urban areas. The earthquake was followed by a succession of typhoons that resulted in slope failures, erosion of slopes and inevitable road loss or undermining in several locations. Erosion has continued with every subsequent typhoon season to the point that slopes regress back on an annual basis, requiring further realignment into the hillsides. Erosion scars immediately beneath the carriageway can plunge almost vertically for 20 m or more, creating highly dangerous conditions for traffic.

Initial hazard assessment

Under a previous contract, landslide hazard mapping had been undertaken using aerial photographs and ground truthing at a scale of 1:10 000. The hazard mapping classified the numerous erosion and landslide embayments within the route corridor according to mechanism of failure and perceived level of threat posed to the road. The mapping formed a useful checklist of possible landslides and slope failure locations.

Desk study and ground reconnaissance

Land surface evaluation techniques applied to the feasibility design comprised aerial photograph interpretation, tabulated inventory and mapping of engineering geology, geomorphology, landslides and erosion at 1:2000 scale. The extent of detail portrayed at this stage was such that it allowed:

- the preferred solution at each site to be defined and its feasibility to be confirmed;
- outline survey and calculation of the likely quantities of retaining structures, slope stabilization and erosion protection works associated with each preferred solution.

Slope stability analyses and outline design of stabilization measures were performed at high-risk sites using input parameters derived from the field mapping and sensitivity analyses for the range of materials, anticipated groundwater and failure types encountered.

A limited amount of subsurface drilling and more detailed topographical surveys were undertaken at sites where reconstruction feasibility was uncertain, or where the derivation of a reliable cost estimate required more detailed information and interpretation.

Detailed design

This was based largely on the confirmation and development of feasibility schedules through the preparation of engineering geological/geomorphological mapping at scales of 1:1000 and 1:2000, together with stereonet analysis and a programme of drilling, trial pitting and cone penetrometer probing. Trial pitting, with close control on material descriptions and classification, proved particularly valuable from a geological, stability and design point of view. The interpretation of mapping and ground investigation data at each site was then used to develop slope stability analysis and the scheduling of road reinstatement and remedial measures. Elsewhere, where slope problems were considered to be less severe, design schedules were based on field judgement using qualitative assessment of engineering geological mapping and ground investigation data alone.

Output review

The design output comprised 1:2000 scale drawings showing the locations of proposed structures and related

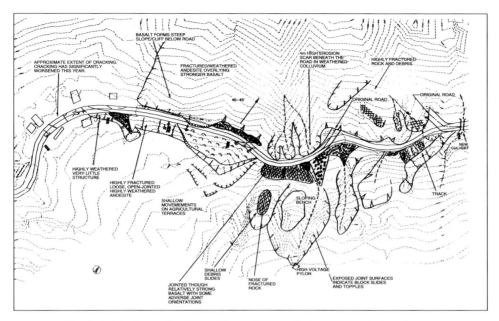

Fig. 3. Original plan (1:2000 scale) showing principal geomorphological features.

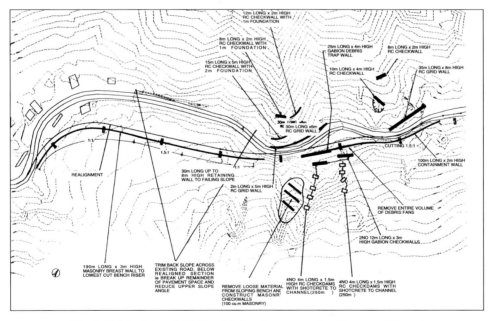

Fig. 4. Original plan (1:2000 scale) showing proposed outline remedial works.

work, together with detailed schedules and annotated site photographs. Figures 3 and 4 illustrate the development of geomorphological and engineering geological mapping into a general arrangement of works scheduled to reinstate the road formation, improve the stability of adjacent slopes and protect slopes and drainage channels from further erosion.

Large-scale engineering geological and geomorphological mapping proved to be the most valuable and mostly widely applied of the land surface evaluation techniques used. The aerial photographs, while useful in examining the wider geomorphology of the road corridor, were in places so adversely affected by scale distortion and shade that they became difficult to use. When compared to engineering geological mapping and trial pitting, subsurface drilling investigations provided relatively little added value to the geotechnical interpretation of the observable shallow surface instability affecting the road. However, drilling did confirm that the low strength of surface materials did not improve significantly with depth, and this general lack of foundation stability was an important factor in the geotechnical decision not to deploy large structural schemes in the approach to road reinstatement and slope stabilization.

Summary

In the case of the design of new road construction, engineering geological and geomorphological evaluation can and should play a major role in the development of design, from initial desk study and route corridor feasibility assessment through to detailed design and the preparation of contract documentation. The use of aerial photograph interpretation and engineering geological/geomorphological mapping through the deployment of relevant personnel directly in the engineering design team generates this facility. In the Nepal study, a different set of documents and inputs to decision making and the design itself would have arisen had land surface evaluation been undertaken by specialists who were not directly involved in design. The advantages of the integrated approach include more cost-effective use of mapping techniques, a closer familiarization with the terrain and its implications for design, and an immediate input of geological and geomorphological data into important design decisions as the project proceeds.

In the case of road rehabilitation, the design relied far more on detailed site appraisal, and used a range of methods for recording this information. Engineering geological and geomorphological mapping at scales of 1:2000 and larger provided an essential means whereby slopes were evaluated qualitatively and a decision made through team discussion over the relative merits of local realignment, widening into the cut or widening on the outside using retaining wall, earthworks design and spoil disposal, etc. The geotechnical design of each
cross-section and the detailed location of the centre-line with respect to cut, fill, topography, foundations and stability was essential to the stability of the formation, its associated structures and the slopes upon which it was founded. Land surface evaluation at the detailed level was therefore essential.

In both the Nepal and Philippines studies, the question remained as to how much information should be recorded if design decisions are to be made more or less instantaneously, as part of a multidisciplinary engineering team. There is certainly a point where the recording of too much detail becomes cost-ineffective, and deciding on the cut-off point is a matter of judgement, combined with a knowledge of what level of detail is required for design. This comes with experience, but the examples given in Figures 1 and 3 serve as a useful illustration from the point of view of a low cost road scheme. Having engineering geological and geotechnical personnel on the project throughout design and implementation allows slopes to be reappraised and more data to be collected, especially as slopes are excavated and the performance of the design can be evaluated. This staffing arrangement is essential when low-cost engineering works are expected to perform under adverse topographical and stability conditions, when the perception of hazard, and a design that leads to an acceptable level of risk, can vary from one wet season to the next.

The geomorphological mapping component used is that described by Brunsden et al. (1975a, b). The degree of detail collected and the manner in which it was applied to design represent a development illustrating how engineering geomorphology can and should be incorporated into the design process.

Did the outcome match the prediction?

Although neither of the road construction and rehabilitation schemes had been implemented at the time of writing, a large rock avalanche in Nepal provides an interesting comparison between outcome and prediction. Figure 2 summarizes the geotechnical hazards recorded or predicted along one section of the Arun Access Road river route. At Tumbetar a combination of rock fall and toe erosion was described. The 1:2000 scale engineering geomorphological plans prepared for this section of the alignment recorded rock fall deposits, adverse jointing and dilation in the cliffs above, and consequently a horizontal alignment was plotted that provided a safety margin between the road and the toe of the cliffs. A channel was also to be cut in the terrace between the road and the cliff to provide drainage and extra rock trap capacity. Conventional survey techniques were used to monitor rock dilation in the cliffs adjacent to the road.

In July 1996 a rock avalanche took place with an estimated volume of 750 000 m^3 and covered the terrace at this location. The trigger mechanism for this avalanche is unknown, although the earlier rock dilation was thought to have arisen through seismic disturbance during an earthquake in 1988. The occurrence of the rock avalanche in this location illustrates the value of land surface evaluation in identifying potential slope hazards, but their magnitude and timing, or frequency of recurrence, remain extremely difficult to judge prior to the event.

Acknowledgements. The author would like to thank the Department of Roads, His Majesty's Government of Nepal, and the Department of Public Works and Highways, Republic of the Philippines for permission to publish this paper. Bob Weekes of Scott Wilson provided the engineering input to field survey and design in Nepal, while Jonathan Hart, also of Scott Wilson, provided some of the geological information from the Philippines.

References

BRUNSDEN, D., DOORNKAMP, J. C., FOOKES, P. G., JONES, D. K. C. & KELLY, J. M. H. 1975a. Large scale geomorphological mapping and highway engineering design. *Quarterly Journal of Engineering Geology*, **8**, 227–253.

BRUNSDEN, D., DOORNKAMP, J. C. FOOKES, P. G., JONES, D. K. C. & KELLY, J. M. H. 1975b. Geomorphological mapping techniques in highway engineering. *Journal of the Institution of Highway Engineer*, **22**, 12, 35-Q1.

FOOKES, P. G., SWEENEY, M. MANBY, C. N. D. & MARTIN, R. P. 1985. Geological and geotechnical engineering aspects of low-cost roads in mountainous terrain. *Engineering Geology*, **21**, 1–152.

TRANSPORT RESEARCH LABORATORY. 1997. *Principles of low cost road engineering in mountainous regions*. Overseas Road Note 16, Transport Research Laboratory, Crowthorne, UK.

Terrain hazard around the Ok Tedi copper mine, Papua New Guinea

G. J. Hearn[1], R. Blong[2] & G. Humphreys[3]

[1] Scott Wilson Kirkpatrick & Co Ltd, Basingstoke, Hampshire, UK
[2] Natural Hazards, Research Centre, MacQuarie University, Australia
[3] School of Earth Sciences, Macquarie University, Australia

Rationale

Located in the remote Star Mountains of Western Province, Papua New Guinea, the Ok Tedi open-cut copper mine operates under conditions of particular topographical and geotechnical adversity. Slopes are frequently steep, rocks are highly fractured and weakened due to tectonics and weathering, and rainfall is world record-breaking, with annual totals exceeding 10 m in the vicinity of the mine (Jones & Maconochie 1990). The mine is planned to occupy an area of 2 km diameter with a depth of 500 m at an altitude of 1800 m a.s.l. The mine supply corridor, along which power, food and other essential materials are brought in and copper concentrate is piped out, is 160 km long, and is itself at risk from slope, flooding and erosional hazards.

The threat posed by these hazards to the operation of the mine and the viability of its supply corridor has been recognized and evaluated by the geotechnical department of Ok Tedi Mining Ltd (OTML) since the project's inception. However, the severity and proximity of this risk were perhaps not fully realized until 1989 when a rock avalanche involving perhaps as much as $70 \times 10^6 \, m^3$ of material occurred within a very short distance of the mine office. The collapse of the mountainside was dramatic and was considered to have been responsible for the seismic event recorded at the time of failure in Port Moresby, approximately 800 km away. Although the infrastructure of the mine was not directly affected by this failure, it served as a catalyst for a multidisciplinary geotechnical study of slope hazards in the mine operational area, which focused on techniques of land surface evaluation rather than intensive subsurface investigation. These techniques are reviewed below and included regional geological and geomorphological mapping, hazard and risk mapping, engineering geological inventory, detailed geomorphological mapping, and risk assessment of landslide and avalanche runout. As such, this study probably represents one of the most intensive and wide ranging applications of land surface evaluation techniques for engineering schemes in remote mountain terrain, and is perhaps unique in the mining sector.

The area model

A regional geomorphological map with a scale of 1:10 000 covering an area of 120 km² was prepared using published and project-derived geological mapping, aerial photograph interpretation and substantial ground checking in accessible areas. One of the main outputs from this mapping exercise was the differentiation between areas and landforms dominated by structural control and those areas where large landslides and colluvial deposits predominated. Individual landform features identified on these maps included ridge and ravine topography, karst landforms and cavity collapse, landslide scars and failure deposits (including rock avalanche flow structures), igneous dykes, gully erosion, sackung rock relaxation features and fault lineations.

The geomorphological map enabled patterns of landform development, drainage and slope stability to be recorded and interpreted in conjunction with maps and cross-sections showing the underlying geology prepared by others in the multidisciplinary study team. This mapping programme led to the conclusion that most of the valley floor landforms are less than 10 000 years in age – youthful even in comparison with other areas of the Papua New Guinea highlands. Furthermore, it was concluded that the most important influence on recent geomorphological evolution in the mine area had been the occurrence of large rock avalanches, a conclusion that clearly had important implications for the geotechnical management of the mine and its infrastructure. Thus, the 1:10 000 scale regional geomorphological and geological mapping became the area model and provided a framework within which other geomorphological techniques were applied, and specific land surface risk and management issues were addressed.

Techniques used

Hazard and risk mapping

The findings of the regional mapping study, combined with the observable slope failures around the mine, led

From: GRIFFITHS, J. S. (ed.) *Land Surface Evaluation for Engineering Practice*. Geological Society, London, Engineering Geology Special Publications, **18**, 143–149. 0267-9914/01/$15.00 © The Geological Society of London 2001.

to the decision by OTML to fund a landslide hazard and risk mapping study (Fookes 1997). This study was split into two parts: the first was at a scale of 1:100 000 and investigated the magnitude and frequency of large landslides; the second was at 1:10 000 scale and concentrated on individual slopes and drainage channels around the mine and the service corridor. The former was based on aerial photograph interpretation, the regional mapping and the development of a database of landslide events, including mechanism, geology, volume and approximate date or age. The latter was based more on field inventory using conventional engineering geological methods and hazard and risk assessment.

Magnitude–frequency assessment

Combining historical records of landslides and avalanches, together with landslide features mapped from aerial photographs taken since 1969, a database of 79 large landslides was established. Volumes were estimated from aerial photographs and field survey, and an age was assigned to each failure on the basis of historical and community knowledge, landform freshness and vegetation patterns, aerial photograph evidence for the more recent failures and ^{14}C dating of organic matter retrieved from prehistoric avalanches. In addition, the consequences of these failures were categorized according to the geomorphological effects described in Table 1. The largest rock avalanche in the area occurred approximately 8800 years ago with a volume of 7×10^9 m^3 and, on the basis of landforms mapped from aerial photographs and topographical maps, deposited material to a depth of 360 m in the Ok Tedi river. One of the smaller avalanches mapped was only 20×10^6 m^3 by comparison and occurred in AD 1977, raising the level of the Ok Tedi river bed by only 2–3 m, at a distance of 30 km from the avalanche source.

Table 1. *Magnitude of future landslides estimated from the 100 year landslide record*

Recurrence interval (years)	30	100	1000
Probability of occurrence in a 30 year period	0.64	0.25	0.03
Volume ($\times 10^6$ m^3)	70	120	>200?
Area (km^2)	3.2	4.5	>5?
Geomorphological consequences*	5	7	8?

* 1, Cliff or slope retreat; aggradation of lower hillslopes; 2, movement/erosion/removal of older colluvium; 3, stream bed aggradation and/or stream bed incision; 4, stream bank trimming and landslide initiation; 5, channel migration; 6, creation of landslide dams, blockage of rivers; 7, truncation of streams and/or ridges, drainage derangement; 8, catchment boundaries modified by landslides.

The practical outcome of this study was the development of a model of geomorphological risk posed by landslides and rock avalanches in the mine area with the conclusion that the evolution of the landscape over the last 15 000 years has been dominated by catastrophic slope failures followed by periods of 'recovery' from them. The observation that an avalanche with a volume similar to that at Vancouver Ridge in 1989 can be expected to occur every 30 years on average in the mapped area is one of the most significant outcomes in terms of mine planning. The comparatively high frequency of small- to medium-sized landslides (up to 100×10^6 m^3) has clear implications for the day-to-day running of the mine, and emphasized the need for a more detailed review of the mine area and its service corridor.

Hazard and risk mapping

The establishment of a hazard and risk categorization for the entire mine operating area was accomplished through systematic field inventory assisted by the regional geological mapping and a review of drilling logs and other geotechnical records held at the mine. Aerial photographs at 1:30 000 scale were examined during this study, but the dense forest cover over much of the area obscured ground detail except in the immediate vicinity of the mine and parts of the service corridor where it had been cleared.

The study area was divided into 245 geomorphological units of average size 40 000 m^2 and shown at a scale of 1:10 000. Rock slope and soil slope inventories were completed for each unit, detailing engineering geological, slope geometry, drainage and stability data. These data were used to compile hazard and risk classifications for each unit according to the procedure described in Hearn (1995*a*). The principal geomorphology of the study area was mapped from aerial photographs and field survey, and was combined with the regional geological mapping to define geomorphological domains in which hazard and risk were summarized for easier review by OTML. In addition to providing the basis for the mapping procedure, the inventories and the associated documentation provided recommendations for remedial measures, further investigation or monitoring schemes, and the timescales over which they should be applied.

Large-scale geomorphological mapping

Following the failure of Vancouver Ridge, there were two main sites of geotechnical concern. The first was Vancouver Ridge itself, where the confirmation of the causes, mechanisms and sequence of failure, and an assessment of the stability of the remainder of the ridge, were considered essential. The second was New York Ridge, immediately to the east.

The structural geological mapping and three-dimensional modelling of Vancouver Ridge remains unpublished, though an overview of these studies is provided by Read & Maconochie (1992). Instability on New York Ridge had been monitored for some time by OTML, the concern being that further deterioration and eventual slope failure could jeopardize the stability of the primary crusher, the ore processing facilities, the mine waste dump and the mine offices, all of which are located in the vicinity.

Site 1: Vancouver Ridge

Methods of land surface evaluation applied to the geomorphological evaluation of this area included conventional stereo aerial photograph interpretation (Fig. 1), sequential plotting of slope erosion, slope instability and channel incision assessment (from photography taken by helicopter at various intervals since the commencement of the project), and 1:2500 and 1:10 000 scale geomorphological field mapping. The extent to which field mapping could safely and effectively be undertaken was limited by the precipitous and highly disturbed slopes that formed the avalanche scarp, and the unpredictability of dense cloud and heavy rain that could very rapidly create extremely dangerous conditions for personnel and air rescue.

The cause of the avalanche could be explained by reference to the following geological factors: an adversely orientated underlying thrust plane, a hanging wall drop fault which could have formed the avalanche release surface, and the low strength of the crushed limestone that constituted the ridge itself. The triggering mechanism was stream incision beneath the waste dump.

It was critical to the geomorphological interpretation of the sequence of failure that large-scale slope dilation was detectable from site photographs taken prior to the collapse of the ridge. A combination of aerial photographic evidence together with geomorphological mapping of the scarp, failed blocks and avalanche flow structures, enabled the sequence of slope failure and deposition to be defined. Furthermore, this mapping information provided the basis for evaluating the flow dynamics of the avalanche: most notably, a travel distance of 3500 m at a velocity of 70–80 km/h.

Site 2: New York Ridge

By contrast, the geotechnical evaluation of New York Ridge relied more heavily on structural geological mapping and detailed geomorphological field mapping over a much larger area (Hearn 1995b). Field mapping was undertaken at a scale of 1:2500 over the entire 4 km^2, and locally at 1:500 in the vicinity of the primary crusher.

With the aid of drillhole information, the geological mapping led to the conclusion that the thrust which had underlain Vancouver Ridge had been partly intruded by the diorite that forms the bulk of New York Ridge, and thus the potential for widespread failure was lower.

On the basis of slope steepness, stream undercutting, past failure morphology and visible signs of recent or active movement, the geomorphological mapping established a classification of slopes in terms of failure potential and potential failure volume, and allowed important conclusions regarding the stability of critical areas to be made. However, even with detailed 1:500 scale geomorphological mapping, augmented by additional drilling and geotechnical analysis, it was not possible to say that slope failure would not progress to the point that it would eventually compromise the stability of the nearby primary crusher within 20 years. An alternative crusher site was therefore investigated by OTML.

Further mapping applications

Despite the inability of the multidisciplinary study to provide a clear conclusion over the stability of the crusher, it was decided by OTML that a combination of structural geological and geomorphological mapping, augmented by drillhole information, was the best way forward in the assessment of slope stability and risk, both in the mine operational area as a whole, and in other key areas. Land surface evaluation techniques were, therefore, further utilized with the following objectives:

1. to assess the overall stability of a proposed site for a large in-pit mobile crusher;
2. to assess the stability of the mine access road in critical areas;
3. to determine the risk posed by potential landslides on the slopes above the mining township.

Although aerial photographs were used to provide general interpretative overviews of landform patterns, the density of rainforest canopy throughout much of the study area, outside the immediate vicinity of the mine and its supply corridor, precluded detailed ground interpretation. Detailed geomorphological studies had to rely almost entirely, therefore, on field mapping. This observation tends to confirm the conclusions of Fookes et al. (1991) regarding earlier aerial photograph interpretations of a landslide site in the study area.

The first two assessments combined 1:2500 scale structural geological and geomorphological mapping with drillhole information to develop conclusions that formed the basis of geotechnical decision making (Hearn 1995b). An illustration of the mapping output is shown in Figure 2. The assessment of risk posed by landslide runout onto the terrace occupied by the mining township is also reported in Hearn (1995b) and comprised

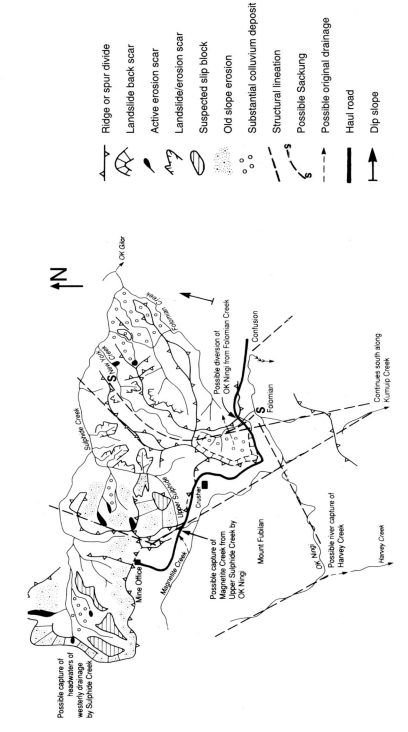

Fig. 1. Aerial photograph interpretation of the Vancouver Ridge–New York Ridge area, Ok Tedi copper mine (original scale 1 : 30 000).

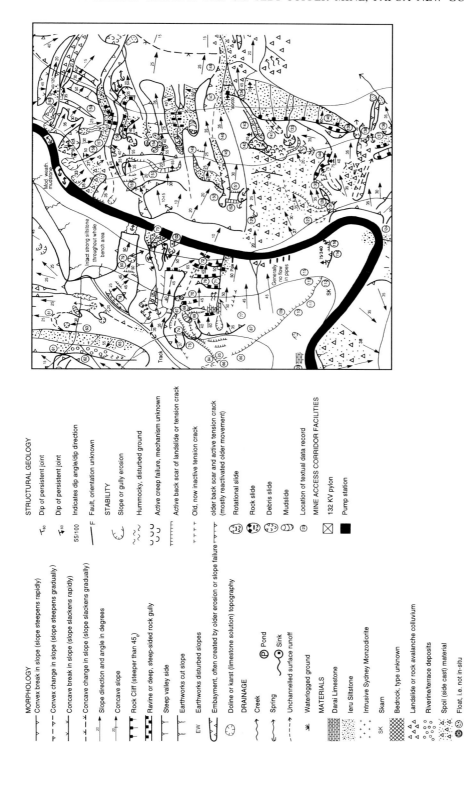

Fig. 2. Example extract of detailed geomorphological mapping for landslide investigation at Parrot's Beak Ridge, Ok Tedi copper mine (original scale 1:2500).

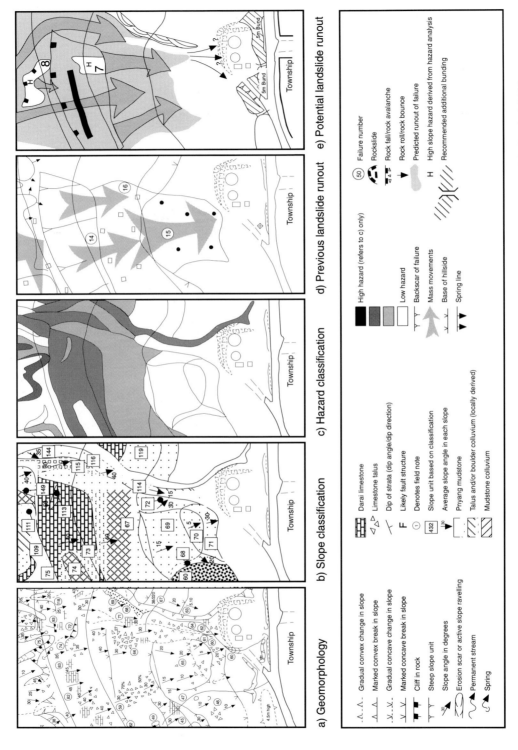

Fig. 3. Extract of landslide hazard and runout mapping above the Ok Tedi township, Papua New Guinea (original scale 1:2500).

detailed geomorphological mapping, landslide hazard assessment and runout modelling over an area of $3 \, km^2$ (Fig. 3). With ground detail obscured by dense rainforest on aerial photographs, the approach adopted relied entirely on 1:2500 scale geomorphological mapping to define slope morphology and identify failure landforms. The mapping was also used to collect data for a formal hazard mapping analysis, together with the development of empirical runout equations to assess the potential for future failures to extend onto the township terrace.

Output

From the consultants' perspective, the land surface evaluation and related techniques applied at Ok Tedi provided both summary and detailed assessment of hazard and risk posed by landslides and related phenomena to mine facilities and operations, and in particular the sites of high value housing and processing plant. This was appreciated not only by the OTML technical staff but also by senior management as a result of a series of in-house seminars. These highlighted the findings in terms of risk awareness, feasible risk minimization options, and an improved understanding of the way the natural environment behaves and reacts to major slope failures and engineering interference.

Eight years on, from OTML's current Geotechnical Superintendent's perspective, 'the multidisciplinary geotechnical study and the associated geohazards work has proved a good investment.' The geohazards work is referenced whenever stability issues occur within the mapped region, and recently the geohazards studies have formed the basis for geotechnical risk assessment in the operational area of the mine. Currently, OTML's

Engineering Services Roads Section and the OTML Mine Engineering–Geotechnical Section are developing contingency plans for the mine access road.

Acknowledgements. The authors would like to thank Ok Tedi Mining Limited (OTML) for the opportunity to undertake the study and for permission to publish this paper. The contribution to the discussion by Mr T. N. Little (current Superintendent Geotechnical at OTML) is gratefully acknowledged. Professor P. G. Fookes acted as review consultant to OTML for the Multidisciplinary Geotechnical Study. Dr R. Mason of CSIRO, Melbourne, Australia, was responsible for the geological mapping described in this paper.

References

FOOKES, P. G., DALE, S. G. & LAND, J. M. 1991. Some observations on a comparative aerial photography interpretation of a landslipped area. *Quarterly Journal of Engineering Geology*, **24**, 249–265.

FOOKES, P. G. 1997. Geology for engineers: the geological model, prediction and performance. *Quarterly Journal of Engineering Geology*, **30**, 293–424.

HEARN, G. J. 1995a. Landslide and erosion hazard mapping at Ok Tedi copper mine, Papua New Guinea. *Quarterly Journal of Engineering Geology*, **28**, 47–60.

HEARN, G. J. 1995b. Engineering geomorphological mapping and open-cast mining in unstable mountains. A case study. *Transactions of the Institution of Mining and Metallurgy, Section A: Minerals Industry*, **104**, A1–A17.

JONES, T. R. P. & MACONOCHIE, A. P. 1990. Twenty five million tonnes of ore and ten metres of rain. Mine Geologists' Conference, Mount Isa, 2–5 October, 159–165.

READ, J. R. & MACONOCHIE, A. P. 1992. The Vancouver Ridge landslide, Ok Tedi Mine, Papua New Guinea. *Sixth International Symposium on Landslides*, Christchurch, New Zealand, 1317–1321.

GIS-based landslide hazard mapping in the Scotland District, Barbados

G. J. Hearn, I. Hodgson & S. Woddy

Scott Wilson Kirkpatrick & Co Ltd, Basingstoke, Hampshire, UK

Rationale

The Scotland District occupies an area of approximately 60 km^2 forming the northeast coastline of Barbados (Fig. 1). The district is geologically and topographically distinct from the rest of Barbados in that the coral limestone cap that covers the remainder of Barbados is absent. The underlying Tertiary rocks exposed in the Scotland District are soft and erodible, and are themselves prone to landsliding and erosion. Land use in the district comprises forest, sugar cane, fruit and vegetable growing and pasture, and the district has high landscape value and eco-tourism potential. Low rise, low density housing has developed over the decades, predominantly along cart tracks and roads with alignments that tend to follow ridge and spur lines in an attempt to avoid potentially unstable side-long ground. However, there are numerous locations where erosion and ground movements have caused significant distress to road pavements, and where foundations to residential buildings have been undermined to the extent that some areas have been abandoned altogether.

The Soil Conservation Unit (Ministry of Agriculture and Rural Development) has been tackling slope instability and erosion problems in the district for over half a century. A number of useful projects have been commissioned to examine and remediate areas of known ground instability, and several schemes involving drainage and earthworks have proved successful. However, realizing that there were several geological, topographical, drainage and land use factors distributed throughout the district that had a role to play in the pattern of ground instability and erosion, the Soil Conservation Unit commissioned a Geographical Information System (GIS)-based project to use remotely sensed and desk study data sources to prepare maps of land degradation. The approach adopted to develop this GIS-based registration, analysis and mapping is described below.

The area model

The geology of the Scotland District is underlain by marls, siliceous mudstones, sandstones with clay–shale interbeds, clays and conglomerates. Folding and faulting are widespread, and consequently local lithological dip orientations are complex and multiple. According to the published geology (Barker & Poole undated) there is a general consistent northeast to east-northeast trend in the larger faults and folds, indicating that the sediments have been deformed by northwesterly orientated compressive forces. Major displacements have taken place with isoclinal folding, overfolding, thrust-faulting, shearing and intrusion of Joe's River Beds into the country rocks as a result of Andean tectonic compression during the Miocene–Pliocene (Barker & Poole undated). These beds are structureless, distorted and slickensided, often containing oil and rafts or fragments of the earlier rocks (see description in Senn (1940) and Fan *et al.* (1996)).

The geomorphology of the Scotland District is defined by the delimiting coral limestone escarpment almost everywhere to the west, and drainage basin topography developed on the Tertiary rocks and recent sediments to the east. The coral limestone escarpment is approximately 33 km long and up to 90 m high. The topography to the west of the cliff slopes gently towards the west, i.e. away from the Scotland District. The gradual recession of the cliff westwards has meant that many valleys draining westward have become progressively beheaded. On the Scotland side of the cliff, and along the cliff itself, there is much evidence of slope failure, with large and numerous detached blocks of limestone deposited in front of the cliff-line and, in some cases, for considerable distances downslope. Tension cracks are also observed behind the cliff-line and are indicative of stress-relief and loss of support as the escarpment has retreated. Springs and seepages are obviously important controls on ground stability.

Techniques used

Methodology

Figure 2 shows the method used to derive maps showing landslide susceptibility, landslide hazard and planning guidance in the Scotland District. Initially, a high specification PC with GIS and image interpretation software was established in the offices of the Soil Conservation

From: GRIFFITHS, J. S. (ed.) *Land Surface Evaluation for Engineering Practice*. Geological Society, London, Engineering Geology Special Publications, **18**, 151–157. 0267-9914/01/$15.00 © The Geological Society of London 2001.

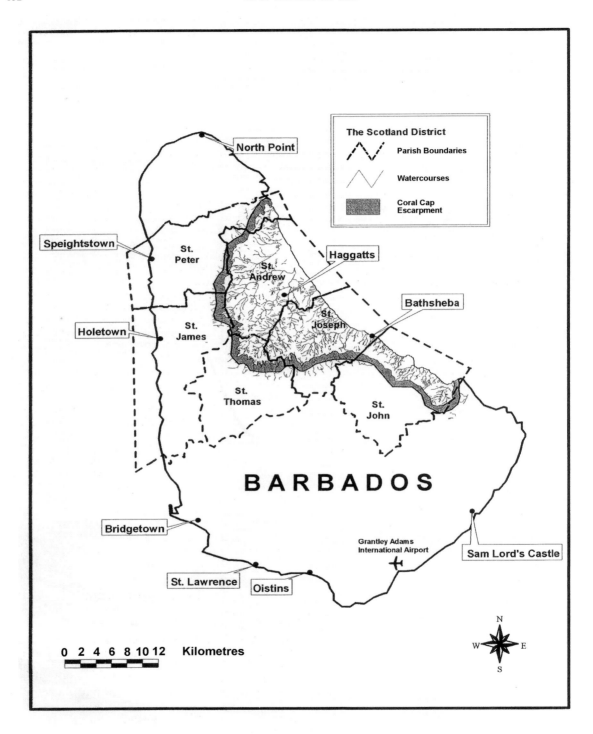

Fig. 1. Study area location.

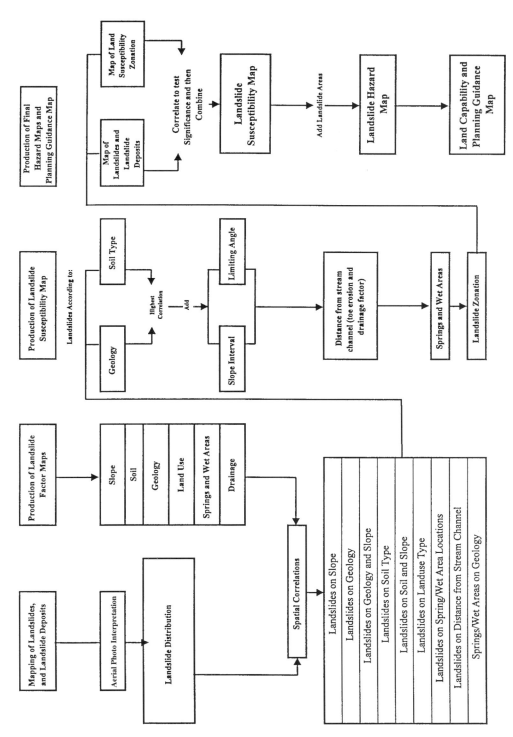

Fig. 2. Activity flow chart for the derivation of susceptibility hazard, and planning guidance mapping outputs.

Unit. Topographical mapping, geological outcrop pattern, agricultural soil types, land use and drainage data sets were obtained from government sources, predominantly in GIS format, and these were co-registered into the Soil Conservation Unit GIS. One of the key activities in this process was the conversion of the different spatial datasets to conform with the underlying National Grid of Barbados, thus allowing compatibility in later spatial analyses. Having established these factor mapping layers, a dataset of landslide events and recorded damage to roads was established from government records. While these damage records were of value in describing the historical record of ground movements, they were insufficient for hazard mapping purposes. Aerial photographs, together with field validation and field observation, became the principal means by which landslide, erosion and drainage data were gathered, interpreted and digitized into the GIS for susceptibility analysis and hazard mapping purposes (Fig. 2).

Aerial photograph interpretation

Black-and-white and colour aerial photographs flown in 1951 and 1991 respectively, and both at approximate scales of 1:10 000, were used systematically to interpret and map landslide and erosion scars throughout the Scotland District. Areas of landslide, areas of creep, and areas of disturbed ground likely to be associated with landslide deposits were recorded onto overlay transparency. Field validation was progressed in parallel with the aerial photograph interpretation. A total of 253 landslides and 313 erosion areas were mapped in this way.

Many of the larger old failures and areas considered to be disturbed by old landslides and landslide materials are located beneath the coral limestone escarpment. Some of these failures appear to have been deep-seated in that they have involved the detachment of large blocks of limestone escarpment. In fact, many of these larger, deep-seated landslides probably involved the detach-

ment of several geological units, as a result of failure through underlying strata. Their distribution is therefore difficult to analyse in any susceptibility mapping as they may bear no simple relationship between stratum, slope angle and failure.

Field observation

In addition to the field validation exercises required for the aerial photograph interpretation, observations were made of natural slope angles at points of ground failure or marginal ground stability in order to identify relationships between material types and limiting angles of stability. The lowest threshold slopes were found to range between 11° for clays, 15° for marls, and 22–28° for sandstones. This range was found to be approximately coincident with relationships derived from the GIS factor analysis described below.

Spatial correlation analysis

The data processing and analysis were mainly carried out using the Geoprocessing Wizard and Spatial Analyst extensions within GIS. The correlation analysis used is based on the chi-squared statistic that examines the observed versus the expected distribution of frequencies according to factor mapping categories. Each GIS factor layer, such as slope or geology, was composed of a number of classes, such as 0–10°, 11–20°, etc., or strata type. The area occupied by each of the classes was expressed as a percentage of the Scotland District area, or the percentage of the area occupied by the sum of the classes, if this did not cover the entire Scotland District. The observed number of landslide areas for the entire Scotland District was then multiplied by each percentage to obtain the *Expected* number of events for each factor class. However, because a given landslide may occupy more than one strata type or more than one land use type for instance, a separate parcel (digitized areal unit) had

Table 1. *Distribution of observed and expected landslide/failure frequencies and areas according to susceptibility categories based on geology and slope*

Landslide susceptibility combined rank	Landslide susceptibility	Total area (m^2)	% Area	Expected no. of failure areas	Observed no of failure areas	O/E	$(O-E)^2/E$	Expected area (m^2)	Observed area (m^2)	O/E
0–3	Very low	14 317 274.89	23.6	221	172	0.78	10.47	1 857 455.59	1 586 878.65	0.85
4–6	Low	14 812 122.52	24.4	228	186	0.82	7.74	1 921 654.78	1 142 516.68	0.59
7–9	Moderate	11 079 088.45	18.2	171	165	0.96	0.21	1 437 348.58	1 538 674	1.07
10–16	High	8 226 618.11	13.5	127	142	1.12	1.77	1 067 282.56	1 549 589.55	1.45
>16	Very high	12 296 875.42	20.2	190	271	1.43	34.53	1 595 338.51	2 061 421.15	1.29
Total				936	936		54.72	7 879 080.03	7 879 080.03	

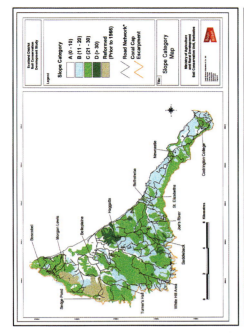

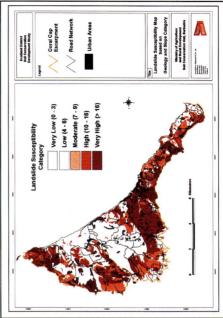

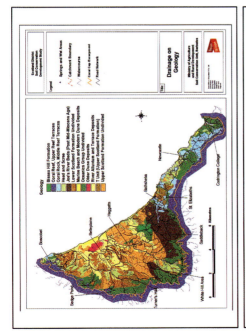

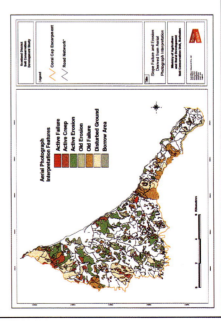

Fig. 3. Illustrations of GIS layers and mapping outputs.

to be created for each. Therefore, one landslide may have been included in the spatial analysis as three parcels in order to account for the fact that it spans three geologies. The *Observed* number of events on each factor class was then counted and the value of *Observed/Expected* (O/E) calculated. This provided a direct indication of the tendency of landslides to cluster on different factor classes. However, this statistic alone did not indicate how significant the control of factor class was on landslide distribution. To determine significance at different levels of probability it was necessary to calculate $(O - E)^2/E$ for each factor class and then sum these values to arrive at a chi-squared value.

Despite the need to use frequency data in the statistical analysis, it is often the areal coverage of landsliding or erosion, rather than the number of events, which provides the best indication of the relative susceptibility of a given factor class to failure, as one large failure on a given factor class is considered to be more significant than a large number of very small failures on the same class. Therefore, once correlation significance had been established on the basis of frequency distribution, it was the O/E values for landslide area that were used in most of the susceptibility ranking of factor classes.

Results

Geology and slope combined were highly correlated with the distribution of landslide areas. However, the relative susceptibility of different rock types was found to be, to an extent, dependent on whether they had been analysed in terms of landslide frequency or landslide area coverage. Susceptibility based on landslide frequency is likely to be biased towards those rock types that are associated more with a large number of small slope failures than a small number of large failures. Susceptibility based on area distribution is more representative of single large failures, and this is reflected in the results obtained. The rock types that outcrop close to the coral cap escarpment were found to be associated with the larger failures, while a larger number of smaller failures were recorded on steeper sections of slope elsewhere. Because of this, the relative susceptibilities based on both analyses were combined into the susceptibility mapping.

This distribution of susceptibility was then tested against the mapped distribution of landslide parcels and the chi-squared data shown in Table 1 were obtained. For landslide frequency there is a sensible progression of O/E values for increasing susceptibility, as would be expected, and the chi-squared statistic is 54.72, which is highly significant. It should be noted that this value is in excess of that obtained for landslides on geology alone even though the latter had more than twice the number of degrees of freedom.

None of the other factors analysed, namely soil type, proximity to a spring or drainage channel, or land use category, showed any meaningful relationship with the landslide distribution, and therefore these were not included in the susceptibility analysis.

Outputs

Figure 3 shows the landslide susceptibility map prepared for the Scotland District, based on the geology and slope factor analysis described above. Relative susceptibility mapping was then combined with the distribution of mapped landslide hazards from the aerial photographs. A three-fold categorization of land capability and planning guidance was developed from this mapping output. Similar maps were produced for erosion susceptibility and erosion hazard using the same method. Recommendations for risk assessment and risk management were made in the accompanying report.

Discussion

The development of a GIS has facilitated the storage, co-registration and analysis of spatial data sets. This has enabled a project-derived landslide distribution to be examined in terms of its relationship with potential landslide-controlling factors. The mapping outputs from this study provide a Scotland District-wide overview of relative landslide susceptibility, rather than an indication of absolute instability at any one location. It is important to bear in mind that the study has been based essentially on pre-existing desk study data sets together with project-derived landslide and erosion distributions based on aerial photograph interpretation. The analysis and mapping outputs are reliant on the accuracy and resolution of these data sources, and the GIS application technology should not be allowed to give a false sense of detail to the user of the maps. This issue can only be addressed realistically when more detailed datasets become available, usually with the assistance of field observation and monitoring.

Although essentially a desk study, the procedure has incorporated field observation wherever possible, in order to combine and validate desk study interpretation with observational data. The added value of the GIS approach to this process is that data can be added and analysed as they become available, and the land degradational model updated and enhanced accordingly.

Acknowledgements. The authors would like to thank the Ministry of Agriculture and Rural Development for the opportunity to undertake this study. Desk study data sets were obtained from various government ministries and departments through the Ministry of Agriculture, and the authors would like to thank the Department of Lands and Surveys and the Coastal Zone Management Unit (Ministry of Environment) in particular for their assistance. Mr Glenn Marshall and Mr John Warner, both of the Soil Conservation Unit, provided invaluable assistance as project counterparts.

References

BARKER, L. H. & POOLE, E. G. Undated. *The geology and mineral resource assessment of the island of Barbados. Part 1: The geology of Barbados.*

FAN, CHEN-HUI, ALLISON, R. J. & JONES, M. E., 1996. Weathering effects on the geotechnical properties of argillaceous sediments in tropical environments and their geomorphological implications. *Earth Surface Processes and Landforms,* **21,** 49–66.

SENN, A. 1940. Palaeogene of Barbados and its bearing on the history and structure of the Antillean-Caribbean Region. *Bulletin of the American Association of Petroleum Geologists,* **24,** 1548–1610.

Ground conditions and hazards: Suez City development, Egypt

D. K. C. Jones

Purpose

Rapid engineering geological/geomorphological surveys are sometimes required for extensive urban and industrial developments on tracts of land for which little or no background topographic or geotechnical information is available. In areas unobscured by vegetation, geomorphological mapping based on air-photo interpretation augmented by field checking and detailed ground survey, can quickly provide a robust framework of terrain units of value in:

(1) planning cost-effective site investigations;
(2) interpreting and extrapolating subsurface point data provided by trial pits/ boreholes;

(3) the early identification of problematic locations requiring detailed investigation;
(4) the preliminary identification and delimitation of surface material resources;
(5) establishing a basis for preliminary assessments of ground hazard potential.

The classic example of such requirements arose in the proposal to redevelop and expand dramatically the size of Suez Town (El Suweis) in the aftermath of the October 1973 Yom Kippur War. The proposed new Suez City was to be developed on 85 km² of desert surface that had previously been little investigated and was presumed initially to be relatively uniform and non-problematic (Fig. 1). However, preliminary site

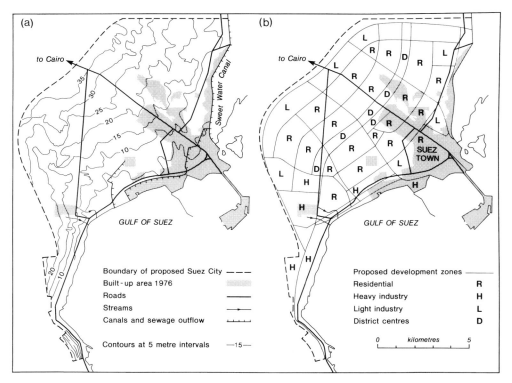

Fig. 1. Maps of Suez, Egypt, showing (**a**) topography and extent of urban area in 1976 and (**b**) proposed layout for a much enlarged Suez City. The sewage works is the square-shaped built-up area immediately to the west of the town; the fertilizer factory is the oblong area further to the west.

From: GRIFFITHS, J. S. (ed.) *Land Surface Evaluation for Engineering Practice*. Geological Society, London, Engineering Geology Special Publications, **18**, 159–169. 0267-9914/01/$15.00 © The Geological Society of London 2001.

investigations revealed unexpected lateral variations in near-surface materials, thereby necessitating a rapid geomorphological survey to establish terrain conditions and to identify the potential for aggressive soils and flash-flood hazard. This was achieved by a team of five specialists in three weeks; details are to be found in Doornkamp *et al.* (1979), Bush *et al.* (1980) and Cooke *et al.* (1982).

The site

Suez Town is located at the northern extremity of the Red Sea where the Suez Canal enters the Gulf of Suez. The proposed urban development was to be on gently sloping and slightly dissected ground (Fig. 1) rising to the west and northwest towards the base of the precipitous Gebel Ataka (871 m) some 3 km beyond the western boundary of the proposed development (Fig. 2A). The relatively uniform surface coloration and insignificant topography were typically deceptive, concealing variations in ground conditions that only began to emerge with the commencement of exploratory subsurface investigations.

Techniques used

Scale and detail of mapping are functions of operator skills, time and the quality of available base materials.

In this case a 1:25 000 scale geomorphological map of the whole of the Suez City site was prepared by interpretation of generally poor quality monochromatic air-photography onto acetate overlays. These were subsequently checked and added to in the field, using procedures and symbolic notation developed in the earlier Bahrain Surface Materials Resources Survey (Doornkamp *et al.* 1979, 1980; Cooke *et al.* 1982) (Fig. 2B). Available air-photography included 1:40 000 scale cover for the whole area and extending northwards and westwards beyond the boundaries of the proposed development; *c.* 1:20 000 scale for the area north of the present town; and some 1:60 000 scale enlargements for the coast and town. All were of some antiquity (1956), which was an advantage as they showed ground form prior to extensive disturbance caused by military activities and more recent urban developments (the value of old maps and photography should never be underestimated). The checked boundaries were then transferred onto a base map created by modifying pre-existing poor quality 1:25 000 scale maps of the area. The resulting materials-orientated geomorphological map and accompanying descriptions of identified units and surface materials could be used at the local level (as in Fig. 2B) and as a synoptic map for the whole development area (Fig. 3).

The geomorphological map was then reinterpreted to yield a surface materials map (Fig. 2C and Fig. 4) which evolved through the incorporation of subsurface data as they became available from the ongoing site

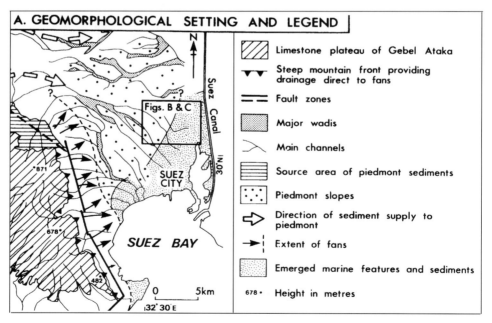

Fig. 2. Rationalization of borehole and trial pit information through a geomorphological and geological interpretation of landforms (Doornkamp *et al.* 1979).

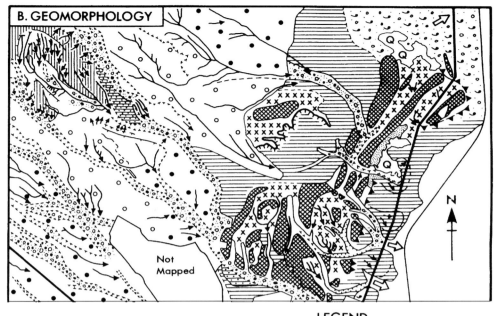

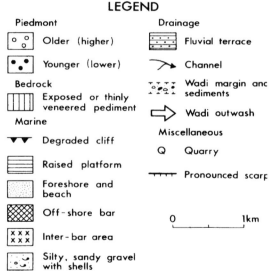

LEGEND

Piedmont

○ ○ ○ Older (higher)

• • • Younger (lower)

Bedrock

▥ Exposed or thinly veneered pediment

Marine

▼▼ Degraded cliff

≡ Raised platform

░ Foreshore and beach

▨ Off-shore bar

x x x / x x x Inter-bar area

⌣ Silty, sandy gravel with shells

Drainage

▦ Fluvial terrace

⤳ Channel

⌣⌣ Wadi margin anc sediments

⇨ Wadi outwash

Miscellaneous

Q Quarry

⊤⊤⊤ Pronounced scarp

0 _____ 1km

Fig. 2. (*continued*).

investigation. Full details of surface geology and superficial deposits are to be found in Bush *et al.* (1980).

The geomorphological map subsequently formed the basis for a preliminary assessment of flood hazard potential, where it was combined with analysis of 1 : 40 000 air photography of the area between Suez City and Gebel Ataka to yield information of wadi catchment areas (Fig. 5). The geomorphological and surface materials maps were also used as the basis for an initial assessment

of aggressive ground conditions. For this, information on surface sediments was combined with:

(i) further air-photo interpretation to distinguish variations in darkness (dampness) tones in order to establish the capillary fringe limit (Fig. 6);

(ii) ground survey of the distribution of damp ground, 'puffy soils' (saline rich), salt efflorescence and depth to water table;

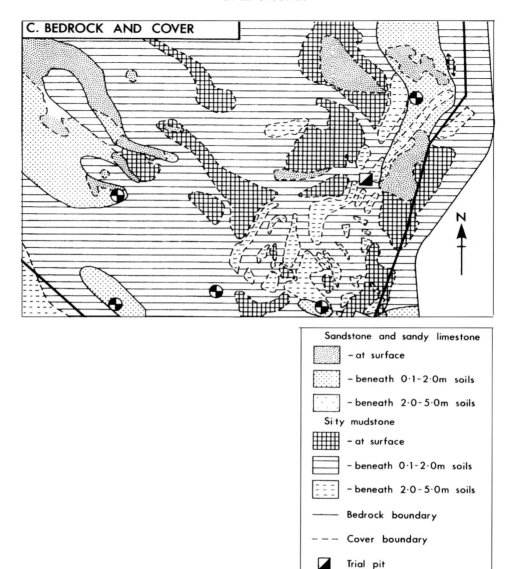

Fig. 2. (*continued*).

(iii) a survey of building damage (Tables 1–3);
(iv) water sampling for simple conductivity testing using a portable meter.

This yielded a preliminary map of salt weathering potential (Fig. 7) that could serve as a basis for a detailed programme of monitoring and chemical analysis.

Conceptual framework

The programme of investigation was underpinned by several fundamental concepts.

1. Near-coastal desert environments often contain a surprisingly diverse range of surface conditions, the existence of which is frequently obscured by the

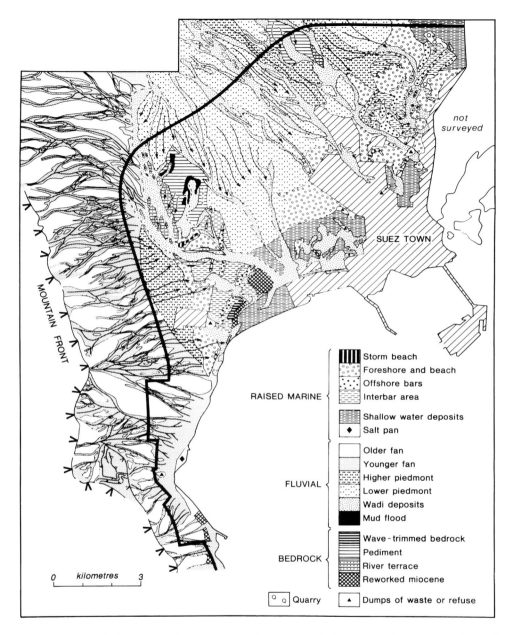

Fig. 3. Geomorphological map of the Suez area produced by air-photo interpretation and ground mapping.

relatively uniform surface coloration and shallow drifts of sand and silt. Morphological/geomorphological mapping carried out by experienced personnel facilitates the identification and delimitation of landforms which, in turn, provide invaluable information on genesis, rela-

tive age and the spatial extent of distinctive suites of surficial materials. Although such information can be built up gradually from site-specific investigations (trial pits/boreholes), it is more efficient to obtain a general appreciation at an early stage of a project through

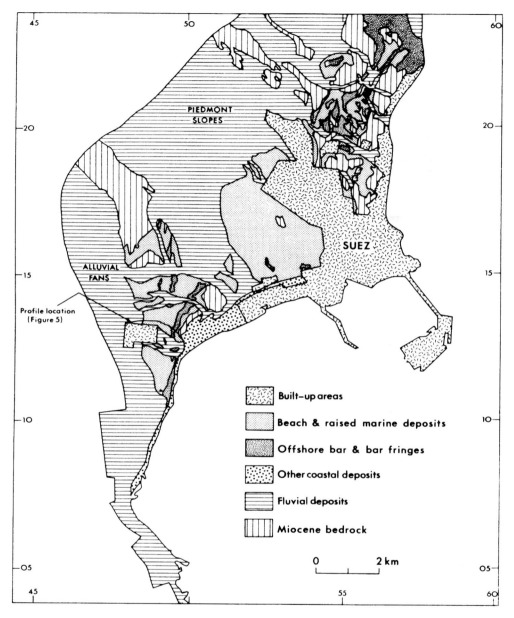

Fig. 4. Main landform units of the Suez area (Bush *et al.* 1980).

geomorphological mapping. The established framework of landforms and deposits can then be used as the basis for planning costly site investigations and extrapolating subsurface data obtained in boreholes, trenches, pits and natural exposures. Thus using the general picture to provide the basis for examining the specific (location/problem) is invariably preferable to using the results of specific studies to establish the general picture, most especially because potentially problematic sites can often be identified at an early stage.

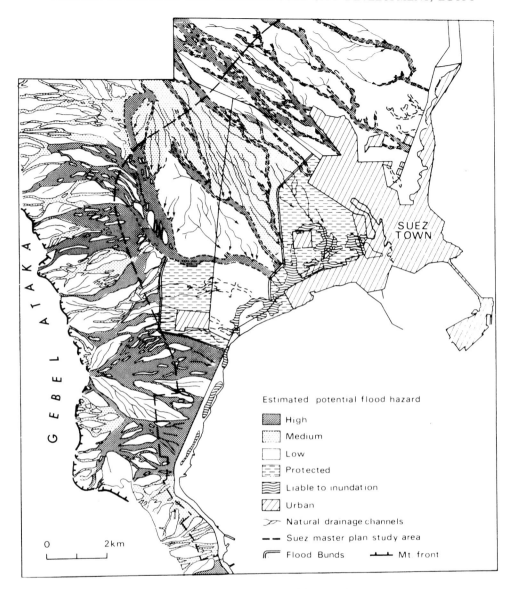

Fig. 5. Flood hazard potential, Suez City.

2. Establishing a general (synoptic) appreciation of terrain conditions (Fig. 3) allows the feasibility of generalized plans to be evaluated at an early stage.

3. The identification of ground hazards requires the careful assessment of all available types of evidence. For example, morphological evidence of flash flooding (i.e. wadi appearance) is often deceptive, as signs of the infrequent occurrence of major discharges can be rapidly obscured by aeolian sediments. Scrutiny of sedimentary deposits, historical records and personal accounts are all invaluable sources of information that need to be integrated together to create a realistic appraisal of hazard potential.

4. The potential for hazard impact at a particular site due to dynamic or transport hazard (e.g. flood, debris flow, avalanche, ash fall, etc.) cannot be ascertained by reference to the observed conditions at the site alone but must involve consideration of all areas that could

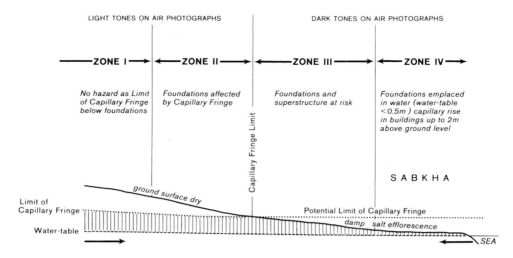

Fig. 6. Methodology used for estimating aggressive ground hazard, Suez City.

yield material to threaten the site in question. This crucial relationship between *site* and *situation* means that site-specific or localized studies are often inadequate determinants of hazard potential. What are required are broader investigations of all areas that could contribute to the threat potential of a specific location: in this instance, all wadi systems flowing through the development zone up to their catchment boundaries on Gebel Ataka.

Survey results

The mapping programme, despite being hampered by the lack of good quality aerial photographs and restrictions on access arising from military activity, quickly revealed that the apparently uniform desert surface actually consisted of three distinct groups of landforms (Fig. 4):

(i)   low relief features developed on Miocene bedrock (sandstones and siltstones) or bedrock thinly veneered with surficial sediments;

(ii) emergent marine deposits (beaches, bars, shell banks, deeper water sediments) with an inland cliffed boundary at 17.2 m (dated at *c*. 44 000 BP (Bush *et al.* 1980)) representing intermittently falling late Quaternary sea-levels in the Gulf of Suez;

(iii) low-angled fans and piedmont slopes shallowly dissected by wadis, all produced by drainage off Gebel Ataka. The survey divided both fans and piedmont slopes into two groups on the basis of age and found evidence for the faulting of bedrock against recent gravels, thereby indicating the area to be seismically active.

The materials-orientated geomorphological map and accompanying sediment descriptions readily explained the observed variations in site investigation findings and allowed the redesign of the site investigation programme to increase cost-effectiveness.

The preliminary assessment of flood hazard was accomplished by examining wadi channel form, channel sediments, channel slope, drainage area and reported occurrence of flash-floods. Although records of recent flash-floods were limited to two occurrences (in 1970 and 1971), the existence of extensive flood bunds (Fig. 5) pointed to the recognition of the hazard and both local inhabitants and the army were aware of the threat. The identification of fine-grained sediments, scour zones, bank erosion and gullying all indicated the occurrence of flood events, but the crucial evidence with respect to scale was the recognition of two small areas of remnant mudflood deposits (Fig. 3). These had survived from an event of sufficient volume/velocity to rise over a 2 m terrace bluff onto a higher surface. The resulting qualitative assessment of flood hazard into *low*, *medium* and *high* relative threat (Fig. 5) was based on catchment areas identified on 1:40 000 scale air-photography, and clearly indicated that the original development plan had failed to take sufficient note of flash-flood hazard potential, thereby necessitating some redesign and the incorporation of flood protection measures. The map (Fig. 5) served as an excellent planning base for such developments and also provided an ideal framework for subsequent monitoring and modelling programmes.

The rapid investigation of the groundwater–salt system yielded maps of depth to water table, groundwater contours and conductivity values, which were combined to yield a groundwater hazard intensity map (Fig. 7) using a four-fold zonal division based on the relationship

Table 1. *Classifications used in survey of building damage by weathering: Suez, Egypt*

Category or measure	Description
I.	**Type of construction**
A	Load-bearing stone walls throughout sometimes rendered, often to 75 cm above ground
B	Load-bearing stone under brick walls often rendered throughout, sometimes with concrete upper floors through walls. Stone extends to 3 m above ground
C	Concrete frame, brick infill, usually on stone base 0–0.5 m above ground
D	Suezi type, commonly mixed stone and brick load-bearing walls with a varying proportion of walling as clay-bonded rubble within light timber cages
E	Mixed construction other than Suezi
II.	**Degree of weathering**
1	Barely perceptible efflorescence
2	Distinct discoloration but no visible surface damage
3–4	Heavy discoloration combined with inception of decay by spalling of rendering along lower edges, slight retraction of mortar in joints and the like
5–6	Widespread loss of rendering, 25% missing and some loosened. Substantial mortar loss in joints or spalling away of base stone and brickwork
7–8	Rendering largely gone, mortar lost from lower joints, about 1 cm of base stone and brickwork spalled away
9–10	Structural breakdown of wall leading towards settlement of living areas sufficient to curtail their utility
III.	**State of construction**
0	Uninhabitable for structural reasons
1–2	Long-neglected crumbling structure, probably best demolished
3–4	Neglected and generally weakened structure, probably capable or restoration
5–6	Stable and generally serviceable structure in average state of repair and decoration
7–8	High quality specification solid and well finished structure with apparent indefinite life
9–10	High quality specification design specifically for Qalzam conditions (a fully effective damp proof course would be obligatory)

Table 2. *Relation of degree of weathering to age and type of construction: Suez, Egypt*

Age and type of construction	Degree of weathering											Totals
	Low									High		
	–	1	2	3	4	5	6	7	8	9	10	
Over 15 years												
Stone	–	1	6	4	4	5	2	3	3	1	–	29
Brick, stone base	–	5	13	13	7	13	7	4	1	–	–	63
Brick, no stone base	–	–	3	4	2	4	2	–	–	–	–	15
Concrete frame	1	3	3	–	3	5	–	1	1	–	–	17
Suezi	1	6	5	7	2	17	5	1	2	1	–	47
Mixed	–	–	2	3	–	2	2	–	1	–	–	10
Subtotal	**2**	**15**	**32**	**31**	**18**	**46**	**18**	**9**	**8**	**2**	**–**	**181**
3 to 15 years												
Stone	1	1	1	–	–	2	–	–	–	–	–	5
Brick, stone base	4	8	14	16	16	12	11	1	1	–	–	83
Brick, no stone base	3	4	3	4	1	4	–	–	2	–	–	21
Concrete frame	2	4	6	10	7	3	4	–	–	–	–	36
Mixed	–	–	1	–	–	–	–	–	–	–	–	1
Subtotal	**10**	**17**	**25**	**30**	**24**	**21**	**15**	**1**	**3**	**–**	**–**	**146**
Less than 3 years												
Brick, stone base	–	1	–	–	–	–	–	–	–	–	–	1
Concrete frame	5	7	6	1	–	–	–	–	–	–	–	19
Subtotal	5	8	6	1	–	–	–	–	–	–	–	20
Totals	**17**	**40**	**63**	**62**	**42**	**67**	**33**	**10**	**11**	**2**	**–**	**347**

Table 3. *Relation of degree of weathering to state of construction: Suez, Egypt*

Age and type of construction	Degree of weathering											Totals
	Low									High		
	–	1	2	3	4	5	6	7	8	9	10	
Derelict –	–	4	4	5	–	3	2	–	1	–	–	19
1	1	2	2	8	–	14	6	2	1	–	–	36
2	–	3	3	6	3	6	6	4	4	–	–	35
3	5	5	13	15	16	18	4	2	3	–	–	81
4	–	3	9	5	7	10	10	–	–	–	–	44
5	8	13	22	17	15	14	4	2	1	2	–	98
6	2	8	5	4	–	1	1	–	1	–	–	22
7	–	2	3	1	–	1	–	–	–	–	–	7
8	1	–	2	1	1	–	–	–	–	–	–	5
9	–	–	–	–	–	–	–	–	–	–	–	
Sound 10	–	–	–	–	–	–	–	–	–	–	–	–
Summary <3	1	9	9	19	3	23	14	6	6	–	–	90
3 to 5	13	21	44	37	38	42	18	4	4	2	–	223
>5	3	10	10	6	1	2	1	–	1	–	–	34
Totals	**17**	**40**	**63**	**62**	**42**	**67**	**33**	**10**	**11**	**2**	**–**	**347**

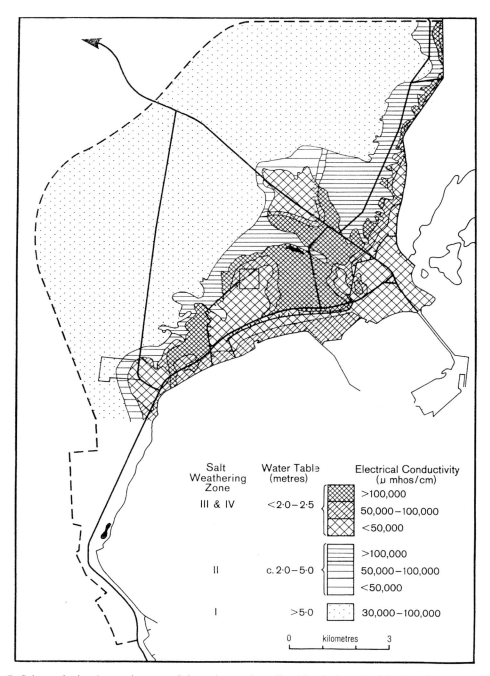

Fig. 7. Salt weathering (aggressive ground) hazard map, Suez City (electrical conductivity: 1 μmho = 1 siemen (S)).

between the topographic surface, the capillary fringe and the water table (Fig. 6) (see Cooke *et al.* (1982) for full discussion). The important points to emerge from the study were:

(i)   the extremely shallow depth of the water table along the coastal belt and associated potential for capillary rise in buildings;

(ii)   the huge range of conductivity values recorded ($4.6–388 \times 10^3 \, \mu S \, cm^{-1}$) and the high values in strongly evaporating zones;

(iii) extensive areas of relatively low conductivity (less than Red Sea values, i.e. $55–60 \times 10^3 \, \mu S \, cm^{-1}$) in the west (near the fertilizer factory), in the centre of the area focusing on the sewage works and in the east beneath Suez Town, adjacent to the Sweet Water Canal and under Feisal City (the north-western extension of Suez Town) (Fig. 7).

These last mentioned freshwater caps or mounds are the products of influent seepage from water pipes, drains, effluent channels and surface irrigation and clearly point to the dynamic nature of the water–salt system following urbanization, thereby indicating the need for monitoring and modelling during the development of near-coastal locations.

Similar applications

Extensive geomorphological mapping based on the interpretation of remotely sensed imagery (satellite or airborne) with checking/extension by ground survey undertaken by suitably skilled personnel, has wide application where there exists the need quickly to understand the spatial distribution of surface materials and potential ground hazards. The level of detail achieved will vary with the scale of the project, quality of available information, nature of terrain, extent and character of vegetation cover, quality and suitability of available imagery, skill of operators and the time and finance available. Dramatic results can be achieved quickly in desert environments whereas surveys of difficult forested terrain are much more labour intensive. Either way, the resulting synoptic picture of surface conditions provides a sound framework for subsequent detailed investigations, especially when used in conjunction with topographic and geological information. It is invariably better to start with the generalized picture of surface conditions (i.e. the context) and then to focus in on specific sites/problems, than to attempt progressively to generalize a context by extrapolation away from scattered points of detailed knowledge as they become available. The same argument has even more force in the case of the assessment of dynamic or transport hazards, where establishing the context of a site (i.e. its situation) is vital to the assessment of hazard potential (threat).

References

BUSH, P., COOKE, R. U., BRUNSDEN, D., DOORNKAMP, J. C. & JONES, D. K. C. 1980. Geology and geomorphology of the Suez city region, Egypt. *Journal of Arid Environments*, **3**, 265–281.

COOKE, R. U., BRUNSDEN, D., DOORNKAMP, J. C. & JONES, D. K. C. 1982. *Urban Geomorphology in Drylands*. Oxford University Press, Oxford.

DOORNKAMP, J. C., BRUNSDEN, D., JONES, D. K. C., COOKE, R. U. & BUSH, P. R. 1979. Rapid geomorphological assessments for engineering. *Quarterly Journal of Engineering Geology*, **12**, 189–214.

DOORNKAMP, J. C., BRUNSDEN, D. & JONES, D. K. C. 1980. *Geology, Geomorphology and Pedology of Bahrain*. Geobooks, Norwich.

Blowing sand and dust hazard, Tabuk, Saudi Arabia

D. K. C. Jones

Rationale

Blowing sand and dust are increasingly common hazards associated with dryland economic development, largely because of disruption of desert soils by human activity (Cooke *et al.* 1982; Cooke & Doornkamp 1990). Blowing sand has long been recognized as a problem; it has clearly identifiable sources (blowouts, scour zones) and sinks (drifts, dunes) and is amenable to a range of management techniques once the problem has been assessed objectively (Cooke *et al.* 1982, 1993). The 'dustification' problem, by contrast, has come to prominence more recently. It is a more complex process involving greater transport distances, and results in a very wide range of adverse impacts including abrasion to paintwork, damage to machinery, contamination of houses, food and water, electrical short-circuits, disruption to radio signals, killing of plants, suffocation of livestock, the spreading of disease, disruption to transport, and general reduction in environmental quality and amenity. The different modes of transport – saltation for sand and suspension for dust – means that while defensive measures may be employed against sand, they are less effective in the case of dust where the emphasis must be placed on curbing the production of fine-grained sediment from identified source areas.

The problem

The strategic town of Tabuk in NW Saudi Arabia was experiencing sand and dust movements of increasing magnitude and frequency. These were having adverse affects on the quality of urban life and military operations because of two fundamental processes: (i) rapid urban expansion in an extremely arid (28.4 mm annual rainfall) environment; and (ii) the increasingly widespread disturbance of the desert surface around the town due to vehicle movements, materials extraction, construction and agriculture. In order to draw up an effective management plan to curb the growing impact of these hazards, a rapid geomorphological investigation (three specialists for three weeks) was undertaken in 1982 to provide an objective assessment of the problem. The study focused on the classic geomorphological task of relating adverse conditions at a particular location (*site*)

to factors/processes arising from elsewhere (*situation*); full details can be found in Jones *et al.* (1986).

The site

Tabuk is located on the western side of a sparsely vegetated basin of interior drainage, the lowest parts of which contain several playas. The basin has a complex low-relief terrain of bedrock pediments, thinly veneered bedrock surfaces and extensive stone pavements at varying elevations, separated by minor scarps and bluffs and cut by shallowly incised wadis. Towards the margins of the basin the amounts of exposed bedrock progressively increase, so that pavements give way to rounded and pinnacled hills cut in near-horizontal sandstone, which become organized into blocks separated by large, sandy-floored wadis. At the western edge of the basin these blocks become higher and more extensive, with a relative relief of over 100 m, separated by well-defined wadis partly filled by sand.

The conceptual model

The fundamental model underpinning the Tabuk study was that the assessment of the blowing sand and dust hazard could be achieved by relating estimations of the *erosivity* of the wind to identified patterns of *erodibility* of the desert surface. Thus the strength and duration of winds capable of transporting sand and/or dust over Tabuk had to be related to the nature of the surfaces over which the winds blow, in terms of their potential to contribute sediment capable of being entrained. The transformation of wind roses into sand and dust roses therefore merely represents the first basic step in the assessment; the potential for aeolian transport then has to be related to the actual landscape in order to estimate the relative combined strength of erosivity and erodibility. This process not only provides a clear indication of the directions from which aeolian transport is most pronounced, thereby providing a basis for planning defensive measures, but also allows the identification of those landform units acting as actual or potential major contributors to sediment transport. This provides a basis for hazard mitigation through land management.

From: Griffiths, J. S. (ed.) *Land Surface Evaluation for Engineering Practice*. Geological Society, London, Engineering Geology Special Publications, **18**, 171–180. 0267-9914/01/$15.00 © The Geological Society of London 2001.

Techniques used

The objective assessment of sand and dust hazard at a specific location requires the integration of four main lines of enquiry:

- evidence for the movement of sand and dust;
- historical data on sand and dust transport episodes;
- identification of conditions suitable for the transport of sand and dust;
- identification of source areas for sand and dust.

Once sediment source and aeolian transport parameters have been established, it should be possible to predict the magnitude and frequency characteristics of the hazards and to propose targeted management strategies.

To achieve these objectives ideally requires the integration of four main methods of data acquisition (Cooke *et al.* 1982):

- analysis of remotely sensed imagery;
- meteorological records;
- geomorphological survey;
- process monitoring.

In this particular case only the first three were possible because of the brief duration of the study.

Model production

Analysis of remotely sensed imagery is important because it can give a broad-scale or regional perspective, as well as provide a basis for mapping. In the Tabuk study, a LANDSAT colour composite of bands 4, 5 and 7 was used to establish the regional setting of the town and to determine the general (prevailing) direction of sand drift. In addition, a 1:50 000 scale monochromatic air-photo mosaic was used to develop a 1:50 000 scale geomorphological map covering an area of 150 km² centred on the town (Fig. 1). This map was checked in the field and concentrated on the identification and delimitation of landform units rather than morphological features, especially in areas of exposed bedrock.

Geomorphological survey was then used to establish evidence for sand and dust transport, in terms of erosional and depositional features, both within the town

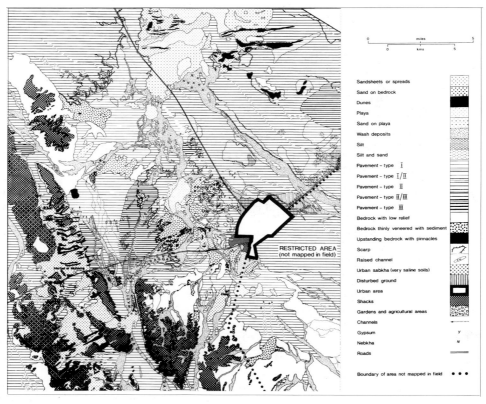

Fig. 1. Reproduction of 1:50 000 scale geomorphological map of Tabuk (from Jones *et al.* 1986).

and in the surrounding desert, and to prove the geo-morphological map. The key refinement was the qualitative assessment of all landform units in terms of their sand and dust drift potential, i.e. the extent to which these units would generate sand/dust if disturbed by human activity. This was achieved by disturbing the surface of units by walking, digging and driving, and resulted in a five-fold classification of stone pavements, with Type I displaying the greatest vulnerability and highest potential to yield entrainable sediments, and Type III the greatest stability (Fig. 1). The final map was then reinterpreted in terms of the sand and dust drift potentials of the recognized units (Figs 2 and 3).

The analysis of limited available meteorological data sought to identify diurnal and monthly patterns of wind speed, and the relationship of higher wind speeds (above

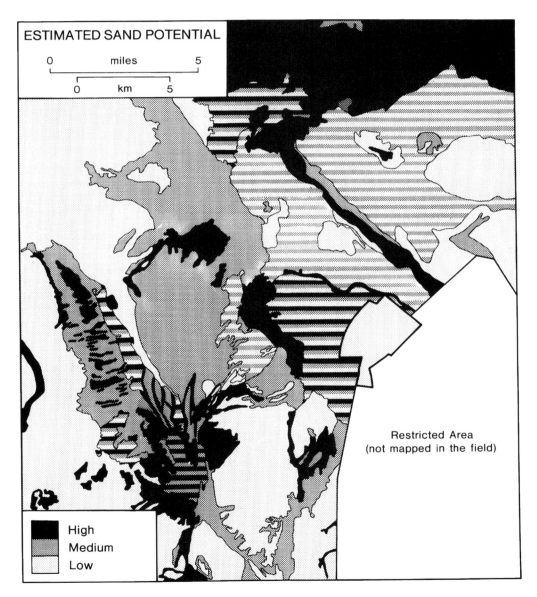

Fig. 2. Map of estimated sand drift potential (Jones *et al.* 1986).

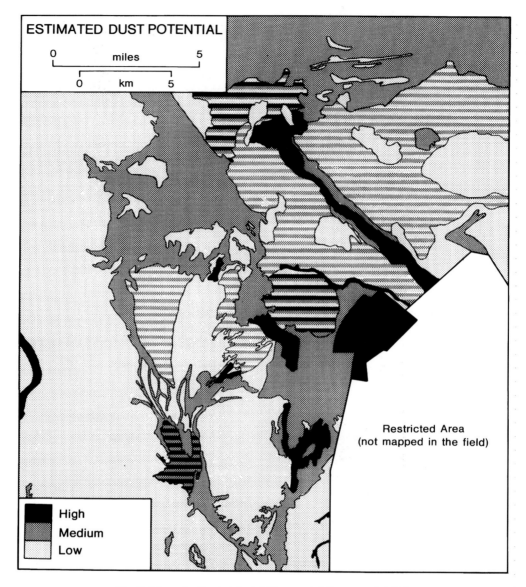

Fig. 3. Map of estimated dust drift potential (Jones *et al.* 1986).

11 knots which is the entrainment velocity for sand) with wind direction and with reduced visibility. These data were then used to calculate the potential drift of sand and dust by month and by direction, leading to the production of 'sand rose' and 'dust rose' diagrams (Fig. 4).

The production of the 'sand rose' diagram was based on Fryberger's (1979) method which uses the formula:

$$Q(V - v_\mathrm{t})V^2 t$$

where: Q = proportional amount of sand drift, V = average wind velocity at 10 m, v_t = impact threshold wind velocity and t = duration of wind.

to estimate sand movement, assuming a threshold entrainment velocity of 12 knots. It is preferable to use a long run of data but in the Tabuk study only two years (1980 and 1981) of records were available. It should be noted that the resulting sand rose assumes unlimited supply of entrainable sand.

SAND ROSE

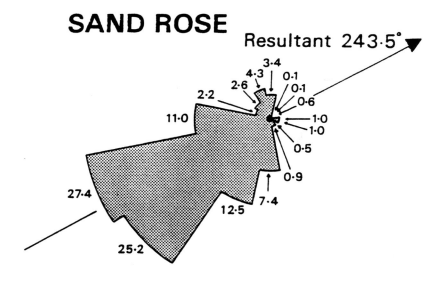

DUST ROSE

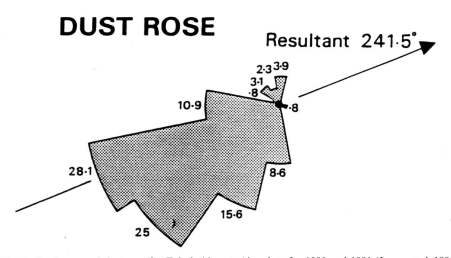

Fig. 4. Sand rose and dust rose for Tabuk Airport using data for 1980 and 1981 (Jones *et al.* 1986).

Calculation of a dust rose is more problematic as the entrainment of dust can occur at a range of wind speeds above 12 knots depending on ground surface characteristics and the presence or absence of the ballistic impact of saltating sand. This problem was overcome by relating dust transport to wind data associated with poor visibility (<3000 m), once it was ascertained that mist and fog were rare. The vast majority of low visibility conditions were found to be associated with wind speeds of over 20 knots and the resulting 'dust rose' was generally similar to that constructed for sand.

Inspection of the LANDSAT image revealed an extensive source of mobile sand some 40 km to the southwest of Tabuk, from which emanated well-developed belts of sand streaming northwestwards across the topographic grain of the bedrock uplands towards the Tabuk basin. Analysis of the orientation of these streamlines and major linear dunes (Fig. 5) revealed dominant sand drift directions of 241.5° across Tabuk and 244.5° slightly to the north of the town, before backing to between 216.5° and 220.4° in the north. The results gave substance to the evidence gained from observations of

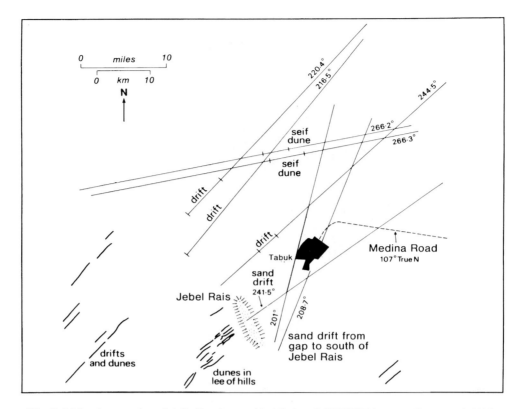

Fig. 5. Major dunes and sand drift directions as identified on LANDSAT imagery (Jones *et al.* 1986).

sand drifts and abrasion features within the town and its immediate environs.

Surprisingly, the 1:50 000 air-photo mosaics did not show the existence of the major streamlines, thereby emphasizing the benefits of employing remotely sensed imagery of different types and scales. The aerial photographs did, however, provide a clear indication of the extent of surface sand and the existence of dunes and drifts, as well as providing an invaluable base for geomorphological mapping (the existence of stereoscopic pairs would have greatly assisted this process). The resultant map (Fig. 1) was based on standard methods developed previously for use in desert areas (Cooke *et al.* 1982) but adapted to the particular needs of the investigation, most especially in terms of the division into 18 surface types and the special focus on pavement type and stability.

The reinterpretation of the 1:50 000 scale geomorphological map to identify potential contributory sources of aeolian sand and dust was focused on the western side of the town. This followed the evidence of sand streams on LANDSAT imagery (Fig. 5) and a pre-

liminary assessment of available wind data that revealed the dominance of winds from the SW quadrant. A qualitative three-fold classification of low, intermediate and high was employed, based on field examination of units. However, in many cases it was found that the potential for landform units to supply sand or dust varied dramatically depending on the nature and extent of human disturbance, resulting in the three intermediate categories low–medium, medium–high and low–high. These maps (Figs 2 and 3) clearly reveal the existence of significant contributory areas close to the town and therefore vulnerable to disturbance.

Meteorological data were only available for Tabuk Airport, immediately southeast of the town, and were of limited duration. Nevertheless, it proved possible to show that significant winds (>12 knots or about the entrainment velocity of sand) occurred for about 8% of the time, mainly in the period 10am to 2am with a pronounced peak in the late afternoon (3pm to 9pm), and that the windiest period was January to May in terms of both wind strength and the duration of strong winds. A wind rose was constructed for the airport and then recalculated to produce a sand rose or sand movement rose (Fig. 4), whose arms are proportional to the amount of sand that could be moved by winds from different directions using Fryberger's (1979) method (see earlier). The resulting 16-arm sand rose based on the two years of available data (1980 and 1981) revealed that 83.5% of sand drift potential was achieved by winds from 169–281°, with easterly winds (011–169°) contributing a meagre 4.2%. The resultant drift direction of 243.5° was remarkably similar to the 241.5° revealed by the examination of LANDSAT.

The production of a dust rose was more problematic (see earlier) and had to be based on wind data associated with visibility recordings of less than 3000 m. The results (Fig. 4) were remarkably similar to those for sand, with 88.2% of dust drift potential from the south and west (169–281°) and a resultant of 241.5°. The next step was to apply the sand and dust roses to the outline of the Tabuk urban perimeter (Fig. 6) in order to yield a map showing the directions from which most sand and dust probably reaches the town. This clearly showed that defences against sand should be

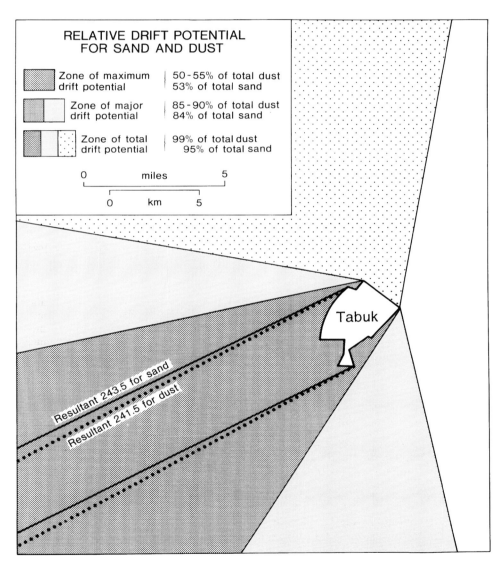

Fig. 6. Sand and dust drift potentials for Tabuk (Jones *et al.* 1986).

focused on the southwestern and western margins of the town.

The assessment of dust hazard was more difficult in the absence of an established procedure and the study had to develop a methodology. The rays of the dust rose (Fig. 4) were applied to the urban perimeter and overlapping segments added together to yield a dust source significance map with values recorded as percentages (Fig. 7). These were indicative of the proportion of aeolian transported dust from a site likely to affect the town (i.e. the higher the value, the greater the likelihood that dust generated by surface disturbance will be blown on to the town). This map clearly emphasized the importance of source areas immediately to the southwest (upwind) of the town. Superimposing this map onto the map of estimated dust drift potential (Fig. 3) and multiplying the percentage values (Fig. 7) by 1 (low), $1\frac{1}{2}$, 2 (medium), $2\frac{1}{2}$ or 3 (high), depending on dust drift potential, resulted in a map showing areas most prominent in the generation of dust if disturbed (Fig. 8).

Use of the model

The survey concluded that the careful management of areas with high values (certainly those greater than 100 and preferably those greater than 50; see Fig. 8) should result in a decrease of the dust problem. Management strategies for such areas could include the provision of metalled roads, the planting of tree belts, restrictions of access onto particularly vulnerable areas, irrigation, restriction of building activity to the southwest, etc.

The general principle of relating site and situation underpins a wide range of engineering geomorphological investigations, so the approach adopted in this study has

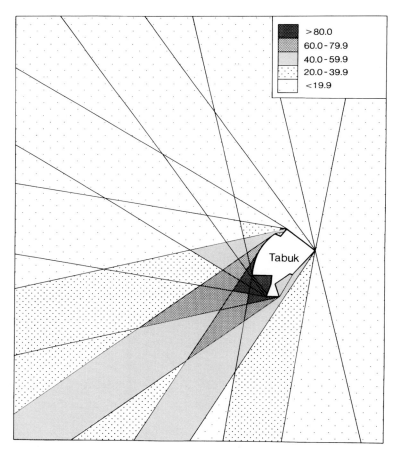

Fig. 7. Dust source potential map for Tabuk (Jones *et al.* 1986).

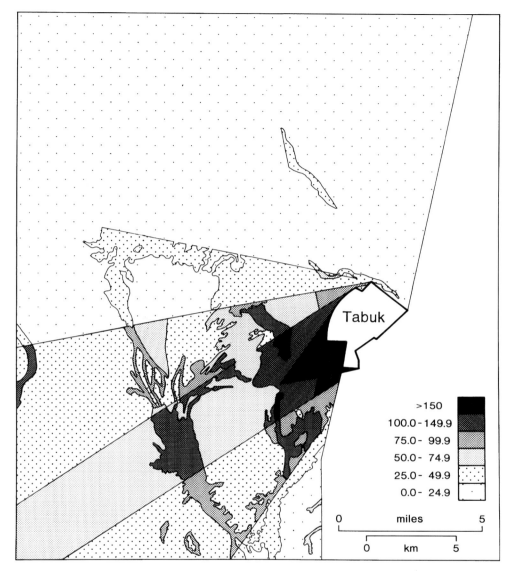

Fig. 8. Dust source significance map for Tabuk showing areas likely to contribute much dust to the town and therefore requiring management (Jones *et al.* 1986).

wide applicability. Similarly, the integration of different kinds of remote sensing, ground survey and other forms of data analysis is of general relevance.

The methodology described is widely applicable to the investigation of sand and dust problems in arid and semi-arid environments. However, it would benefit from the following: the availability of stereoscopic pairs of aerial photographs, longer runs of meteorological data and evidence for changing levels of aeolian transport over time; monitoring of actual aeolian transport episodes using sand and dust traps; monitoring of the behaviour of landform units under different wind conditions; study of the conditions leading to the formation of duststorms; and refinement of the methodology for ground surface classification with respect to the potential for moving sand and dust.

References

COOKE, R. U. & DOORNKAMP, J. C. 1990. *Geomorphology in Environmental Management*. Oxford University Press, Oxford.

COOKE, R. U., BRUNSDEN, D., DOORNKAMP, J. C. & JONES, D. K. C. 1982. *Urban Geomorphology in Dry Lands*. Oxford University Press, Oxford.

COOKE, R. U., WARREN, A. & GOUDIE, A. S. 1993. *Desert Geomorphology*. UCL Press, London.

FRYBERGER, S. G. 1979 Dune forms and wind regime. *In*: MCKEE, E. D. (ed.) *A Study of Global Sand Seas*. US Geological Survey Professional Paper 1052, 137–169.

JONES, D. K. C., COOKE, R. U. & WARREN, A. 1986. Geomorphological investigation, for engineering purposes, of blowing sand and dust hazard. *Quarterly Journal of Engineering Geology*, **19**, 251–270.

Sediment budget analysis for coastal management, west Dorset

E. M. Lee[1] & D. Brunsden[2]

[1] Department of Marine Sciences and Coastal Management, University of Newcastle, Newcastle-upon-Tyne, UK
[2] Emeritus Professor of Physical Geography, King's College, London, UK

Background

The identification and characterization of geomorphological units or systems is particularly important in the coastal zone. Indeed, the consequences of failure to appreciate the physical environment can be more acute on the coast, as rapid, major changes are a reality for land use planning and development. On the coast, it is often more useful to map the landscape in terms of sediment 'cells' (i.e. process units) rather than terrain units (i.e. landform units) as an understanding of the supply and transport of sediment (e.g. sand and shingle) is fundamental to dealing with many shoreline problems. Sediment is circulated in what often can be regarded, for practical purposes, as almost closed cells that are separated by boundaries across which little beach material is transferred. Each cell can be characterized in terms of the inputs, outputs, stores and sinks of sediment.

For example, to understand the development of a beach it is useful to consider it as a store of shingle or sand supplied from source areas on the adjacent coastline or offshore (Fig. 1). Beach building material might be supplied from the seabed, moved onshore by wave energy, or from rivers and eroding cliffs. This material is then redistributed along the shoreline by waves ('longshore drift'), unless prevented by barriers such as headlands or breakwaters. Although longshore drift might be prevented by these barriers, some of the material can still be 'lost' to the system around the seaward end of the barriers or offshore, particularly during large storms.

Sediment inputs and longshore drift are not necessarily constant over time and so it is important to consider the current beach behaviour within the context of the changes that might have occurred over the period of the historical record. Over time, the balance between sediment inputs and outputs (i.e. the sediment budget) within the system will determine whether the beach experiences growth, decline or has remained constant in overall size.

The sediment cell concept is now used as a shoreline management tool in the UK (e.g. MAFF 1995), as it provides a framework for assessing the nature of the interdependence of coastal landforms and assessing the impact of engineering works.

Purpose of the study

Awareness has been growing of the cumulative effects of coastal engineering works on the environment. Particular problems have arisen as a result of the disruption of longshore coarse sediment transport by structures such as harbour breakwaters, piers and groynes. This has led to decline in beach levels and degradation of some sites of Earth science or nature conservation value. It has been recognized, therefore, that there is a need to evaluate the potential effects of coastal engineering works on the adjacent coast by assessing the significance of sediment supply and transport to the sustainability of nearby conservation sites and beaches (Rendel Geotechnics 1995, 1997*a*).

The following example describes the development of a shingle budget and longshore transport model for the eastern part of Lyme Bay, England. The study was undertaken to determine the geomorphological impacts of the current proposed options for coastal defence improvements at West Bay, Dorset on Chesil Beach and the Fleet (Rendel Geotechnics 1997*b*).

The Lyme Bay coast

The shoreline of the eastern part of Lyme Bay (Fig. 2) comprises a series of discontinuous shingle beaches, which show marked changes in character either side of West Bay. To the west, pocket beaches occur at Charmouth, Seatown, Eype and West Beach (West Bay), separated by headlands and backed by high eroding cliffs. To the east, the continuous shingle ridge of Chesil Beach extends for 28 km from East Beach (West Bay) to the Isle of Portland. Most of eastern Lyme Bay is an unprotected coastline, although there are notable exceptions: the seawall at Chiswell, the coastal defences and piers at West Bay and the coastal defences and The Cobb at Lyme Regis.

Chesil Beach is a landform of unique scientific importance, both nationally and internationally (Carr & Blackley 1973, 1974). The beach also performs a critical

From: GRIFFITHS, J. S. (ed.) *Land Surface Evaluation for Engineering Practice*. Geological Society, London, Engineering Geology Special Publications, **18**, 181–187. 0267-9914/01/$15.00 © The Geological Society of London 2001.

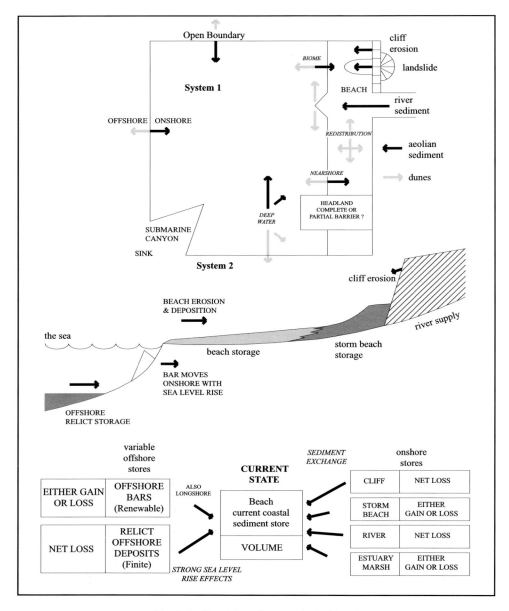

Fig. 1. Sediment transfers on a typical beach.

coast protection and sea defence role for communities such as Chesilton, the Portland Naval base and the important lagoon habitat of the Fleet. It is important, therefore, that proposed coastal defence improvements at nearby West Bay should not affect the operation of coastal processes to the extent that the level of erosion or flood risk along Chesil Beach is increased, or lead to the degradation of the conservation value of the coastline.

Techniques used

The study was based on aerial photograph interpretation and field mapping of the coastal landforms, together with a review of the available published scientific literature (e.g. PhD theses, consultancy reports, historical charts held at the Hydrographic Office, Taunton) and discussions with research scientists who have an expert

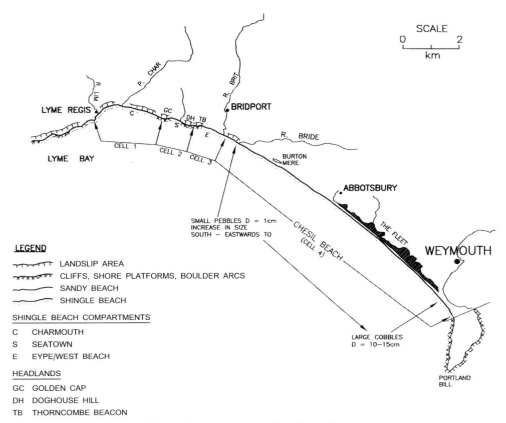

Fig. 2. Shingle transport cells of Lyme Bay.

knowledge of Chesil Beach and sediment movement in Lyme Bay. The study focused on the development of a simple geomorphological model of the coastline between Lyme Regis and the Isle of Portland. This model provided the framework for understanding historical and contemporary changes and for predicting potential future changes associated with coastal defence works at West Bay. The model development involved the following.

- The identification and characterization of shingle transport cells along the coastline. Cell boundaries were defined in the field from known or suspected sediment barriers (e.g. natural headlands or man-made structures).
- The development of shingle budgets for the cells, based on a review of the available literature (notably Bray 1996) and field inspection. Sediment inputs from cliff recession had been previously quantified from a detailed assessment of the geology and geomorphology of the landslide systems along this coastline (Bray 1996). Inputs from rivers had been considered by Rendel Geotechnics (1996). Seabed sediment maps were inspected to identify potential offshore shingle

sources. Important outputs include beach mining operations, with possible volumes of shingle extraction identified from previous studies (e.g. Carr 1985).
- The contemporary evolution of the shingle transport cells and the historical impact of coastal engineering works (such as the West Bay piers) was determined from detailed inspection of historical charts and topographic maps dating back to the end of the eighteenth century.

The geomorphological model

Shingle transport cells

This coastline comprises a sequence of discrete shingle transport 'cells' separated by natural headlands and coastal engineering structures (Fig. 2):

- cell 1 – Lyme Regis to Golden Cap (the western end is defined by The Cobb);
- cell 2 – Golden Cap to Doghouse Hill;
- cell 3 – Doghouse Hill to the West Bay piers;
- cell 4 – the West Bay piers to the Isle of Portland.

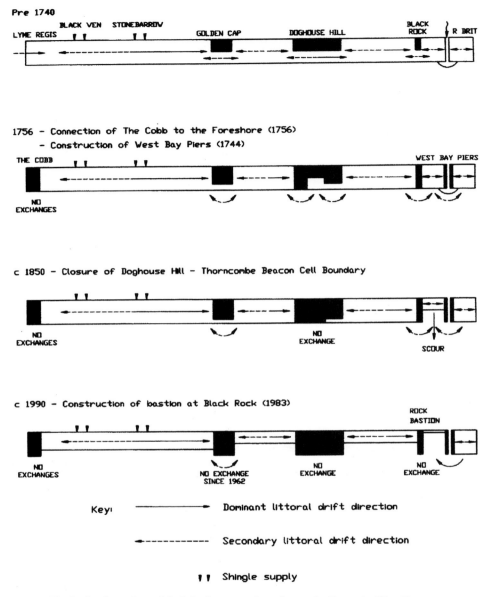

Fig. 3. A schematic model of the fragmentation of coastal cells on the West Dorset coast.

Shingle is moved both eastwards and westwards along this coastline, following the prevailing wind and wave direction at the time. However, as there is a long-term prevalence for strong southwesterly and westerly winds, the overall resultant drift is eastwards. Periods of dominant easterly winds do occur, leading to net drift reversals when the dominant shingle movement is to the west.

Historical chart and photograph evidence suggests that the four discrete shingle transport cells are probably the remnants of what once may have been a continuous shingle beach extending from Lyme Regis to the Isle of Portland. Prior to the construction of The Cobb this beach may have even extended westwards into East Devon. However, the long-term effects of differential

Table 1. *Current shingle budgets for the west Dorset coast*

Cell	Gravel inputs (m³/year; in thousands)					Gravel outputs (m³/year; in thousands)					Balance (input–output)	Beach mining (m³/year in thousands)	Cumulative balance (inc. beach mining) (m³/year in thousands)
	Cliff recession	Rivers	Littoral drift	Onshore transport	Total	Attrition	Littoral drift	Entrapment	Offshore transport	Total			
1. Lyme Regis to Golden Cap	6.18	0.24	0	NS	6.42	1.29	0	0.19	NS	1.48	4.94	0.62 (upto 1956)	4.32
2. Golden Cap to Doghouse Hill	0.09	0	0	NS	0.09	0.08	0	0.05	NS	0.13	−0.04	2.14 (upto 1987)	−2.18
3. Doghouse Hill to West Bay Piers	NS	0	0	NS	NS	0.08	0	0.05	2.7	2.83	−2.83	0.4 (upto 1986)	−3.23
4. West Bay Piers to the Isle of Portland	NS	NS	0	NS	NS	0.2	1.2*	0	NS	1.4	−1.4	16 (upto 1986)	−17.4
Total	6.27	0.24	0	NS	6.51	1.65	1.2	0.29	2.7	5.84	0.67	19.16	−17.25

Modified from Bray (1996).
This condition has operated since 1962 and may continue to do so indefinitely.
* Dredged from West Bay harbour channel.
NS = not significant.

erosion of the more resistant headlands (Golden Cap, Doghouse Hill to Thorncombe Beacon) and the softer rocks in the intervening bays has gradually created the present-day sequence of pocket beaches, which are only episodically linked by shingle transfers. The fragmentation of the continuous beach into discrete pocket beaches may have been a relatively recent phenomenon (Fig. 3).

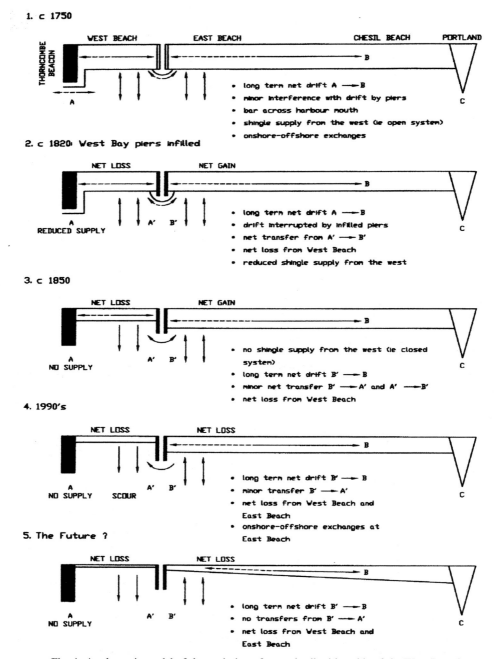

Fig. 4. A schematic model of the evolution of coastal cells either side of the West Bay piers.

Shingle budgets

A contemporary shingle budget for cells 1–4 is presented in Table 1. The inputs into the Lyme Bay shingle transport cells are limited to cliff recession inputs, especially from Black Ven and Stonebarrow. Although the net drift direction is eastwards, the presence of headlands at Golden Cap and Doghouse Hill – Thorncombe Beacon restricts the longshore movement of shingle. Indeed, it is believed that there has been no shingle exchange between the cells since 1820–1850 (cells 2 and 3) and 1962 (cells 1 and 2). The net eastward drift of shingle has been further restricted by the construction and infilling of the West Bay piers, since between 1744 and 1820. Beach mining of gravels in all the cells would have had a significant effect on the shingle budgets until extraction finally ceased in 1987.

Shingle transport, West Bay

Figure 4 provides a schematic representation of the sediment dynamics in cells 3 and 4 highlighting the changes that have occurred since the construction of the West Bay piers in 1744. The 'stages' in this evolutionary model are as follows.

(i) *Pre-1750.* A shingle bar across the mouth of the piers allows the free transfer of material between cells 3 and 4. The continuous shingle beach which probably from Lyme Regis to Portland at this time ensures that the system is open, receiving shingle supplies from the eroding cliffs of East Devon (until The Cobb was connected to the foreshore in 1756), Black Ven and Stonebarrow (in westerly storms) and from Chesil (in easterly storms).

(ii) *Around 1820.* The piers were infilled, further restricting the transfer of material between cells 3 and 4. However, because of the shingle bar at the pier mouth and the high beach levels in cell 3 there remains a net loss of material from cell 3 around the pier heads to cell 4 because of the predominant easterly drift.

(iii) *Around 1850.* The loss of the shingle bar at the pier mouth and the depletion of West Beach between 1820 and 1850 reduces the potential for eastward exchange of sediment between the cells. This condition is exacerbated by the closure of the shingle transport link between cells 2 and 3 with the development of the Doghouse Hill – Thorncombe Beacon headland as a sediment barrier between 1787 and 1850. However, westward exchanges between cells 4 and 3 remain possible because of high shingle levels on East Beach.

(iv) *Late 20th century.* Scour around the west pier and in front of the seawall (built in the late 19th century) had led to severe depletion and set-back of West Beach making further transfers from cell 3

to 4 largely unfeasible. This depletion has been compounded by beach mining operations and the lack of sediment inputs from cell 2 because of the mudslide lobes and boulder aprons below Doghouse Hill, and a rock bastion built at the west end of West Beach.

(v) *The Future.* The net eastward drift in cell 4 will lead to the long-term gradual depletion of East Beach. Periods of westerly drift will no longer lead to the major build-up of material against the east pier and a time will arrive when the westward transfer of material from cell 4 to cell 3 becomes unfeasible.

Potential impacts to Chesil Beach

The shingle budgets and transport models described above provided a framework for a preliminary assessment of the potential impacts on coastal defence improvements at West Bay on Chesil Beach. It was concluded that Chesil Beach is a relict feature with no significant shingle exchanges with West Beach. It is considered unlikely, therefore, that coastal defence improvement works in cell 3 would have a significant impact on the sediment budget and morphology of Chesil Beach. Such works could, however, have significant local impacts which would need to be addressed when considering the suitability of different scheme options.

References

BRAY, M. J. 1996. *Beach budget analysis and shingle transport dynamics in West Dorset.* PhD Thesis, LSE, London.

CARR, A. P. 1985. *Gravel extraction at Cogden Beach.* Report to Dorset County Council, evidence submitted to the Cogden Beach Public Planning Inquiry.

CARR, A. P. & BLACKLEY, M. W. L. 1973. Investigations bearing on the age and development of Chesil Beach, Dorset and the associated area. *Transactions of the Institute of British Geographers*, **58**, 99–111.

CARR, A. P. & BLACKLEY, M. W. L. 1974. Ideas on the origin and development of Chesil Beach, Dorset. *Proceedings of Dorset Natural History and Archaeology Society*, **95**, 9–17.

MINISTRY OF AGRICULTURE, FISHERIES AND FOOD. 1995. *Shoreline Management Plans: a guide for operating authorities.* MAFF Publication **PB2197**.

RENDEL GEOTECHNICS. 1995. *Coastal Planning and Management: a review of earth science information needs.* HMSO, London.

RENDEL GEOTECHNICS. 1996. *Sediment Inputs from Rivers.* Unpublished Report to SCOPAC.

RENDEL GEOTECHNICS. 1997a. *The Investigation and Management of Soft Rock Cliffs in England and Wales.* Report to MAFF.

RENDEL GEOTECHNICS. 1997b. *West Bay Geomorphological Study.* Unpublished Report to West Dorset District Council.

Land use planning in unstable areas: Ventnor, Isle of Wight

E. M. Lee[1] & R. Moore[2]

[1] Department of Marine Sciences and Coastal Management, University of Newcastle,
Newcastle-upon-Tyne, UK
[2] Sir William Halcrow and Partners, Birmingham, UK

Purpose of the study

Recent UK Government guidance has emphasized the need to take account of landslide problems in the land use planning process (DoE 1990, 1996). To assist the implementation of this policy, the then Department of the Environment (DoE) commissioned a number of demonstration projects to develop approaches to assess the potential for landsliding and to identify the best ways of incorporating this information in the planning process.

In the UK there are many situations where historic development has resulted in the concentration of urban development and infrastructure on unstable ground. This is often on a scale such that total avoidance or abandonment are out of the question, as is recourse to large-scale and inordinately expensive engineering solutions (e.g. the Bath area, Lyme Regis, the South Wales valleys, etc. (Jones & Lee 1994)). Under these circumstances, detailed knowledge of slope instability is required so that pragmatic policies can be developed to assist communities to reduce risk. This approach has been pioneered by the detailed study of the Undercliff at Ventnor, on the south coast of the Isle of Wight, England (Lee & Moore 1991; Lee et al. 1991a, b, c; Moore et al. 1991).

The study area

The situation at Ventnor is unusual in that the whole town lies within an ancient landslide complex. The Undercliff has generally been subjected to slow, deep-seated ground movements or creep with less frequent episodic periods of more active movements. Consequently most of the developed area has been affected, resulting in cumulative damage to buildings, roads and services (Lee & Moore 1991; Rendel Geotechnics 1995). Over the last 100 years at least 50 buildings have had to be demolished in Ventnor due to ground movement. The planning problems are, therefore, related to the control of development in those parts of the town which have been shown to be particularly susceptible to ground movement.

The landslides within the Undercliff are developed in Lower and Upper Cretaceous rocks. These consist of sandstones (Lower Greensand) overlain by over 40 m of Gault Clay, then massive cherty sandstones (Upper Greensand) and Chalk. Of particular note is the presence of thin argillaceous layers within the Sandrock of the Lower Greensand, which together with the Gault Clay have a very important influence on the stability and hydrogeology of the area. The geological structure of the Undercliff is relatively simple, with the strata dipping seaward at around 1.5–2° south-southeast. In addition, a NNW–SSE trending synclinal structure is superimposed on the general dip.

Techniques used

The Ventnor study involved assessing the ground behaviour from field mapping and desk study sources alone. The study was directed towards:

(a) determining the nature and extent of the landslide complex;
(b) understanding the past behaviour of separate parts of the landslide system;
(c) formulating a range of management strategies to reduce the impact of future movement.

The methodology used in this study is shown in Figure 1 and highlights the importance of geomorphological mapping to the whole project. The approach involved a thorough review of available records, reports and documents relating to instability, followed by a detailed field investigation comprising geomorphological and geological mapping, photogrammetric analyses, a survey of structural damage caused by ground movement, a land use survey and a review of local building practice (Fig. 1). A search through historical documents, local newspapers from 1855 to 1989, local authority records and published scientific research revealed nearly 200 individual incidents of ground movement over the last two centuries. The various forms of instability that have occurred are summarized in Figure 2.

The results of these investigations provided an understanding of the nature and extent of the landslide system, together with the type, size and frequency of contemporary movements and their impact on the local community. This detailed understanding of ground behaviour was used, in conjunction with knowledge

From: GRIFFITHS, J. S. (ed.) Land Surface Evaluation for Engineering Practice. Geological Society, London, Engineering Geology Special Publications, 18, 189–192. 0267-9914/01/$15.00 © The Geological Society of London 2001.

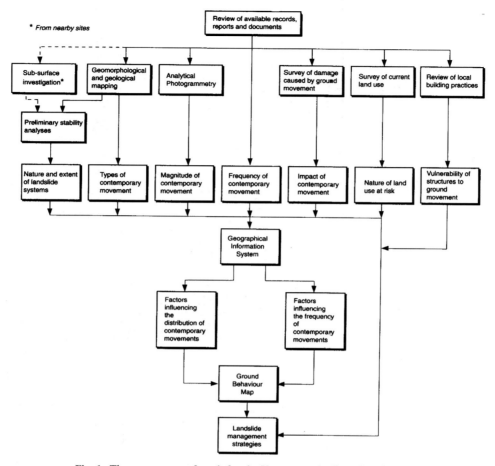

Fig. 1. The programme of work for the Ventnor study (from Lee & Moore 1991).

of the vulnerability to movement of different types of construction and the spatial distribution of property at risk, to formulate a range of management strategies designed to reduce the impact of future movements.

Geomorphological mapping at 1:2500 scale revealed the extent and complexity of the landslides. The Undercliff lies immediately below the Chalk Downs and Upper Greensand bench. From the surface evidence and subsurface work elsewhere in the Undercliff (e.g. Hutchinson *et al.* 1991) the following main features were distinguished:

● a sequence of *compound slides* which occupy a zone of similar breadth in the lower part of the Undercliff;
● *multiple rotational slides* occupy a broad zone in the upper parts of the Undercliff, giving rise to linear benches separated by intermediate scarps: these units mainly comprise back-tilted blocks of Upper Greensand and Chalk;

● in Upper Ventnor, a *graben-like feature* occurs landward of the zone of multiple rotational slides, comprising a 20 m wide subsiding block bounded by parallel fissures, and extends parallel to the coast for over 500 m: this unit exhibits the most serious ground movements recently experienced in the town;
● *mudslides* have developed on the coast where displaced Gault Clay is exposed.

Once the framework of landslide units had been established, it was possible to relate building damage and movement rates to units with known dimensions.

Ground behaviour mapping

One of the main problems with many landslide hazard assessments is that subjective comparisons are made between magnitude and frequency of different processes

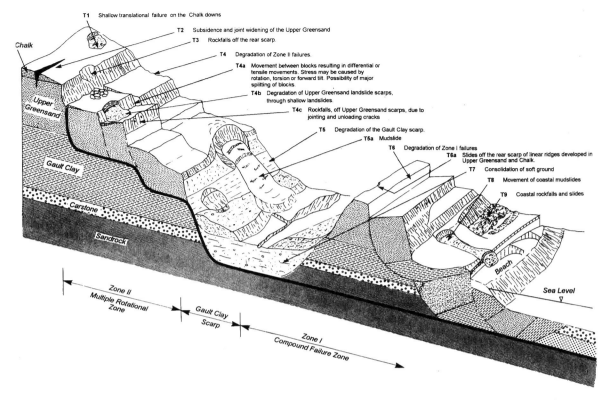

Fig. 2. Types of contemporary ground movement in the Ventnor Undercliff (from Rendel Geotechnics 1995).

within the confines of a simple scale of hazard. For example, Chandler & Hutchinson (1984) devised a preliminary zonation for Ventnor, using four classes of probability (negligible, low, moderate and large) of future movements. The use of such arbitrary subjective scales can cause serious difficulties, as perceptions of what actually constitutes a 'high' probability or a 'small' probability will vary considerably. This can lead to misunderstandings and unnecessary alarm amongst those the maps are intended to help, namely the general public. For these reasons the Ventnor study attempted to avoid subjective scales and has concentrated on analysing and presenting the wealth of available information on movements within the town in a factual and objective manner.

Understanding the geomorphology of the landslide complex at Ventnor proved to be the key to understanding the nature and pattern of contemporary movements. This understanding was used to compile a 1:2500 scale ground behaviour map. This map summarizes the nature, magnitude and frequency of contemporary processes and their impact on the local community, being a synthesis of the following information:

- the nature and extent of individual landslide units which together form the mosaic of landslide features known as the Undercliff at Ventnor;
- the different landslide processes which have operated within the town over the last 200 years;
- the location of ground movement events recorded in the last 200 years;
- the recorded rates of ground movement, over the last 30–100 years;
- the severity of damage to property caused by ground movement;
- the causes of damage to property as a result of ground movement;
- the relationship between past landslide events and antecedent rainfall.

The ground behaviour map demonstrated that the problems resulting from ground movement vary from place to place according to the geomorphological setting. This formed the basis for landslide management strategies that can be applied within the context of a zoning framework that reflects the variations in stability rather than a blanket approach to the problem. In support of the management strategy a 1:2500 scale planning guidance map was produced which related categories of ground behaviour to forward planning and development control. The map indicated that different areas of the landslide complex need to be treated in different ways for both policy formulation and the review of planning applications. Areas were recognized which are likely to be suitable for developments, along with areas which are either subject to significant constraints or mostly unsuitable.

Landslide management

Since the publication of the study the local authority has adopted an Undercliff Landslide Management Strategy, the objectives of which are:

- to reduce the likelihood of future movement by controlling the factors (both natural and man-induced) that cause ground movements;
- to limit the impact of future movement through the adoption of appropriate planning and building controls.

The study identified that considerable benefit can be gained by reducing the frequency and magnitude of ground movement events through engineering works designed to improve the stability of the landslide system. This work would best be directed towards coastal protection to prevent toe erosion or unloading of the landslide complex, groundwater management and improving the infrastructure throughout the Undercliff (some structures have been poorly maintained and neglected to the extent that they have become dangerous) and thereby improve confidence in the area. Potentially the most serious destabilizing factor associated with development in the Undercliff has been the artificial surcharge of groundwater from septic tanks, leaking water pipes, drains, sewers and swimming pools.

References

CHANDLER, M. P. & HUTCHINSON, J. N. 1984. Assessment of relative slide hazard within a large, pre-existing coastal landslide at Ventnor, Isle of Wight. *In: Proceedings of the IVth International Symposium of Landslides*, Toronto, 2, 517–522.

DEPARTMENT OF THE ENVIRONMENT. 1990. *Planning Policy Guidance PPG14: Development on Unstable Land*. HMSO, London.

DEPARTMENT OF THE ENVIRONMENT. 1996. *Planning Policy Guidance note 14 (Annex 1): Landslides and Planning*. HMSO, London.

HUTCHINSON, J. N., BROMHEAD, E. N. & CHANDLER, M. P. 1991. Investigations of landslides at St Catherine's Point, Isle of Wight. *In*: CHANDLER, R. J. (ed.) *Slope Stability Engineering: Developments and Applications*. Thomas Telford, London, 169–179.

JONES, D. K. C. & LEE, E. M. 1994. *Landsliding in Great Britain*. HMSO, London.

LEE, E. M. & MOORE, R. 1991. *Coastal landslip potential assessment: Isle of Wight Undercliff, Ventnor*. Department of the Environment, London.

LEE, E. M., DOORNKAMP, J. C., BRUNSDEN, D. & NOTON, N. H. 1991a. *Ground movement in Ventnor, Isle of Wight*. Department of the Environment, London.

LEE E. M., MOORE, R., BRUNSDEN, D. & SIDDLE, H. J. 1991b. The assessment of ground behaviour at Ventnor, Isle of Wight. *In*: CHANDLER, R. J. (ed.) *Slope Stability Engineering: Developments and Applications*. Thomas Telford, London, 207–212.

LEE, E. M., MOORE, R., BURT, N. & BRUNSDEN, D. 1991c. Strategies for managing the landslide complex at Ventnor, Isle of Wight. *In*: CHANDLER, R. J. (ed.) *Slope Stability Engineering: Development and Applications*. Thomas Telford, London, 219–225.

MOORE, R., LEE, E. M. & NOTON, N. H. 1991. The distribution, frequency and magnitude of ground movements at Ventnor, Isle of Wight. *In*: CHANDLER, R. J. (ed.) *Slope Stability Engineering: Development and Applications*. Thomas Telford, London, 231–236.

RENDEL GEOTECHNICS. 1995. *The Undercliff of the Isle of Wight: a review of ground behaviour*. South Wight Borough Council.

Subsidence map development in an area of abandoned salt mines

E. M. Lee[1] & C. F. Sakalas[2]

[1] Department of Marine Sciences and Coastal Management, University of Newcastle, Newcastle-upon-Tyne, UK
[2] High-Point Rendel, Birmingham, UK

Purpose of the study

Morphological mapping is not generally associated with mining and subsidence, having traditionally been used to define the nature and extent of surface features such as landslides and landforms. However, detailed morphological mapping, linked to subsurface investigation, can provide a preliminary indication of potential subsidence hazards in some areas of abandoned mine workings. This example describes how mapping was carried out in an undeveloped part of the Cheshire saltfield, prior to any subsurface investigation, with the aim of establishing the likely extent to which a number of mines had collapsed.

The site

The site is in the Northwich area of Cheshire, England, and has had a long history of rock salt mining and brine pumping. The site is largely undeveloped, although locally important infrastructure routes cross the area, and it has potential for use as a community woodland and open space area.

At the site, rock salt (halite) occurs in two discrete horizons within the Triassic Mercia Mudstone sequence, separated by around 10 m of marls: the 20–30 m thick Top Bed (at $c.$ 40 m b.g.l.) and the 23–28 m thick Bottom Bed (at $c.$ 75 m b.g.l.). This sequence is overlain by glacial till. Salt was extensively mined at the site from around 1777 to 1933, with the one Top Bed and seven Bottom Bed mines opened over this period, varying in size from 0.8 to 13 ha. Since abandonment there have been both dramatic mine collapses and gradual subsidence, with some mines believed to have remained open or partially open. Details of the mining and subsidence history of the Cheshire area can be found in Calvert (1915) and Wharmby (1987).

In response to a potential purchaser's need to establish the risks and liabilities associated with future mine collapse and subsidence, a preliminary stability assessment was undertaken. This stage of the investigation was very much a low-cost affair. It was not practicable to commence with a conventional borehole investigation. The approach to investigating the site was based upon a number of pragmatic assumptions:

1. at the time of abandonment all the underground workings were unstable: if they had not already collapsed, they may be expected to collapse at some time in the future;
2. the degree of risk and liability across the area will vary with the nature of current and future land uses above the workings and the nature of the works themselves;
3. management responses should reflect the variation in risk across the area, rather than adopting a blanket approach.

Techniques used

The project involved a comprehensive desk study review of the mining and subsidence history of the site, comparison of benchmark levels on different editions of 1:2500 scale Ordnance Survey maps and a survey of structural damage caused by subsidence. Surface morphological mapping was undertaken at a scale of 1:1250 using a tape, compass and hand-held clinometer. The survey was directed towards identifying the subtle (and, in some cases, not so subtle) changes in slope angle and direction that might indicate the spatial extent of mine collapse. It quickly became apparent that the technique was very effective in recording how the natural topography changes in form and begins to 'bend' inwards around abandoned mine workings.

The morphological mapping demonstrated that many of the smaller mines had produced no surface signs of collapse. In addition, a number of larger mines that were believed to have completely collapsed were shown to be, in places, only partly collapsed by comparing the surface morphology with the historical mine plans. On this basis it was possible to recognize four broad groups of mined areas:

From: GRIFFITHS, J. S. (ed.) *Land Surface Evaluation for Engineering Practice*. Geological Society, London, Engineering Geology Special Publications, **18**, 193–195. 0267-9914/01/$15.00 © The Geological Society of London 2001.

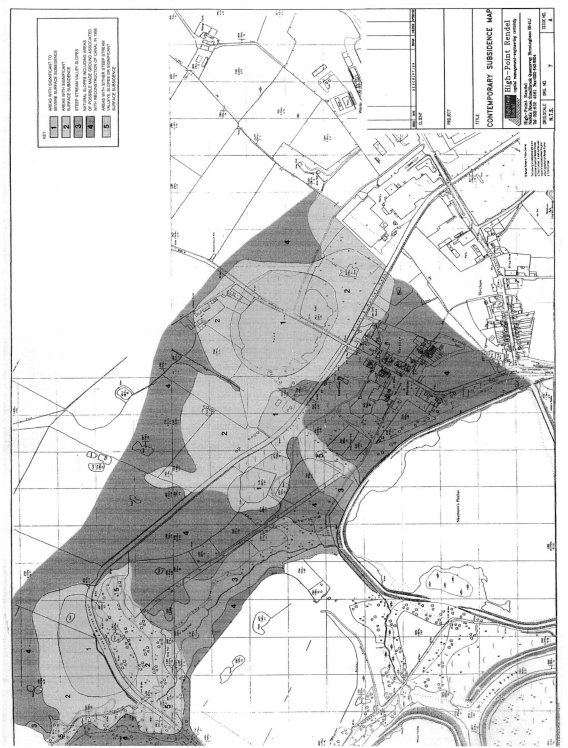

Fig. 1. Contemporary subsidence map, Cheshire.

1. undermined areas which have no surface evidence of collapse: the ground above these mines is characteristically gently sloping southwards;
2. undermined areas which have surface evidence of partial collapse: these areas are characterized by broad, enclosed depressions which go against the 'grain' of the natural topography;
3. undermined areas which have surface evidence of complete collapse, e.g. the major crater which marks the site of the former Top Bed mine;
4. possible collapsed early pits used for extracting rock-salt from the Top Bed; small enclosed depressions on the northern side of the site.

Contemporary subsidence map

A contemporary subsidence map was prepared (Fig. 1) which draws together the varied information about the subsidence history of the site, i.e.:

- recorded evidence of subsidence
- ground movement measurements
- building damage
- morphological mapping
- mine abandonment plans

Bearing in mind that the whole of the area is likely to have been affected by minor ground movements associated with natural subsidence due to natural salt solution or brine pumping, five main land classes were recognized on the basis of the pattern of past mining activity and subsequent ground movement:

1. areas with significant to severe surface subsidence;
2. areas with significant surface subsidence;
3. natural slopes: gently sloping land beyond the area of surface subsidence;
4. stream valley slopes: incised side slopes to stream channels;
5. areas with either steep stream valley slopes or significant surface subsidence.

Areas which have been affected by mine collapse may not necessarily be free of subsidence problems in the future, as these mines may have only partially collapsed.

Subsurface investigation

The contemporary subsidence map, based largely on desk study and morphological mapping, provided a useful preliminary indication of the likely subsidence risks at the site. On this basis the client considered it worthwhile proceeding further with the plans. It was recognized that the potential for future subsidence can only be assessed through subsurface information about present mine conditions and, hence, a further stage of work was carried out. The subsidence map provided a framework for planning the subsurface investigation (including boreholes and ultrasonic surveys). The investigation largely confirmed the general picture inferred from the surface mapping, suggesting that the morphological mapping technique could be a useful preliminary stage in mine subsidence studies in other settings.

References

CALVERT, A. F. 1915. *Salt in Cheshire.*
WHARMBY, P. 1987. *A report on the rock-salt mines and brine shafts in the Cheshire saltfield.* Cheshire County Council.

The design of remedial works to the Dharan–Dhankuta Road, East Nepal

R. P. Martin

Geotechnical Engineering Office, Civil Engineering Department, Government of Hong Kong Special Administration Region, China

Purpose of mapping

Parts of the Dharan–Dhankuta Road in East Nepal were seriously damaged by the Nepal earthquake of 21 August 1988 and subsequently by a locally intense monsoon storm on 12 September 1988. Detailed field mapping was carried out at several of the worst-affected hill sections of the road. The purpose of the mapping was to establish the nature and local extent of earthquake, and subsequent landslide, rockfall and erosion damage as a prerequisite for the design of road remedial works. A full account of the work is given by Roughton & Partners (1988). This is an example of post-construction stage land surface evaluation using the five-stage classification of mountain road projects proposed by Fookes *et al.* (1985).

The site

The Dharan–Dhankuta Road is a 50 km long mountain road traversing the first two ranges of the Low Himalaya of East Nepal (Fig. 1). It was built between 1977 and 1982 under the UK Government's overseas aid programme to the Kingdom of Nepal (Cross 1982).

The epicentre of the earthquake on 21 August 1988 was in the Udayapur District, about 65 km to the west of Dharan (Fig. 1). The focal depth was *c.* 60 km and the surface wave magnitude (Ms) was 6.6 (Dikshit 1991). Three weeks after the earthquake, heavy rain fell over part of the site during monsoon thunderstorms. The 24 hour rainfall recorded in Dharan was *c.* 160 mm, most of which fell within a few hours.

The problem

Serious earthquake damage to the road was confined to the first 26 km from Dharan where the route climbs over the Sangure Ridge. This is a range of hills up to 1500 m in elevation formed in weathered Tertiary Siwalik sandstones/siltstones and Mesozoic quartzites, meta-sandstones and phyllites of the Sangure Series (Fig. 1).

The road was affected by three main types of earthquake damage:

(i) widespread instability in roadside cut slopes and natural rock slopes, especially in the brittle fractured quartzite, ranging from small rockfalls of a few cubic metres to debris slides and rockslides up to several thousand cubic metres;
(ii) extensive cracking in natural slopes mantled with colluvial soils overlying fractured rock;
(iii) significant displacement and occasional complete failure of retaining structures supporting the road formation and cut slopes above the formation.

Although not exceptional for the area, the subsequent monsoon storm triggered further landsliding on slopes already weakened by the earthquake and caused numerous blockages of culverts and other drainage structures.

Emergency clearance operations re-established four-wheel drive, single-lane access along the road within four to six days of the earthquake and storm events. Two-lane access for lorries and buses was re-established along most of the road by mid-October 1988, following the removal of *c.* 65 000 m^3 of landslide and erosion debris. However, the most heavily damaged section of the road, which included two areas of complete failure of the road formation, remained closed to full traffic for several months.

Techniques used

This case study describes the mapping and checklist techniques used for the design of remedial works in the most heavily damaged section located at Karkichhap, north of the Sangure Ridge (Fig. 1). The key sources of information used for the field mapping were copies of 1:500 scale original road alignment plans and 1:100 scale cross-section drawings produced by the design engineers, together with the relevant design reports. As-built record drawings (elevations and cross-sections) at similar scales were also consulted where available, but these were not complete. Also valuable were descriptions of the

From: GRIFFITHS, J. S. (ed.) *Land Surface Evaluation for Engineering Practice*. Geological Society, London, Engineering Geology Special Publications, **18**, 197–204. 0267-9914/01/$15.00 © The Geological Society of London 2001.

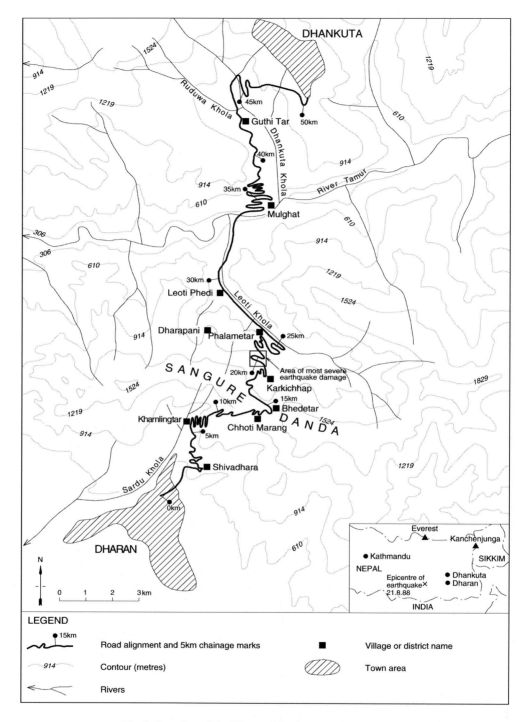

Fig. 1. Location of the Dharan–Dhankuta Road in East Nepal.

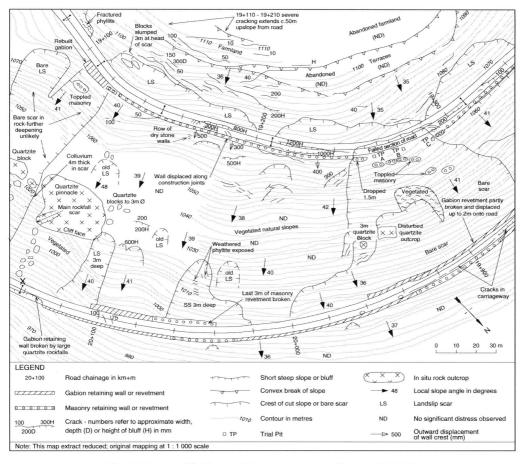

Fig. 2. Extract of field map of damaged section of road at Karkichhap.

natural terrain on geomorphological sketch plans, sections and pro formas compiled during a reconnaissance-stage land surface evaluation of the provisional road alignment made two years prior to the start of construction (Brunsden *et al.* 1981).

Field mapping was carried out using 1:1000 scale, or reduced 1:500 scale, composite road alignment plans with a 2 m contour interval. The mapping included selected elements of geomorphology, surficial geology and damage to the road formation and retaining structures. Figure 2 is an extract of such a field map in the centre of the heavily damaged section of the alignment. Figure 3 shows part of the same area prior to the earthquake and storm damage (**A**), during construction of remedial works (**B**) and after completion of the works (**C**). The mapping was complemented by the use of a standardized checklist (pro forma) to record details of damage to the distressed retaining structures (Fig. 4). The

checklist incorporated additional sketch mapping at a typical scale of 1:200.

Three hand-excavated trial pits were dug along the line of the former road edge in the failed section at Ch 19 + 240–19 + 280, to provide information on suitable founding depths for reconstructed retaining walls (Fig. 2). These were the only subsurface investigations used for the remedial works design. Use of a drilling rig was considered but was ruled out because of difficulties of procurement and access within the works design programme, and the questionable value of point data in such an extensively disturbed area.

Conceptual models

Analysis of the results of the field mapping and checklists allowed conceptual models to be developed for two

Fig. 3. Upper (Ch 19 + 250–19 + 350) and lower (Ch 19 + 800–20 + 000) roadlines (**A**) prior to the earthquake and storm damage (June 1988), (**B**) during construction of remedial works (May 1989) and (**C**) following completion of the works (October 1991). In (**B**) the failed section of retaining wall (see Fig. 6A) has been rebuilt along the upper line while landslide debris still covers part of the lower line. In (**C**) the slope between the roadlines has been reinstated with bioengineering works (terracing and grass/tree planting above a gabion revetment at the toe).

types of slope failure and a mechanism of progressive displacement and collapse of masonry retaining walls.

1. *Rockfall and rockslides in quartzite cliffs.* The earthquake dislodged many individual rock blocks and generated rockslides from two massive quartzite pinnacles in between the hairpinning roadline. The mass of the largest fallen block was estimated at 300 tonnes. Two sections of roadline supported by gabion retaining walls up to 7 m high were largely destroyed by the rockfall impact (Figs 2 and 5). This type of failure was not repeated in the monsoon storm.

2. *Cracking and displacement of soil slopes (natural and cut slopes).* Surface cracking of colluvium and intensely weathered phyllites was visible throughout the area, often associated with shallow debris slides (Fig. 2). Individual cracks were typically 50–300 mm wide, 100–300 mm deep and showed relative vertical displacement of up to 500 mm. The cracks varied in length from a few metres up to a maximum of 50 m (commonly 5–20 m) and in plan shape from straight to sharply concave downslope. Most were aligned parallel or subparallel to the slope contours. The intensity of cracking varied significantly, with the worst distress concentrated in two areas: above Ch 19 + 100–19 + 210 and in between the roadlines from Ch 19 + 320–19 + 430 and 19 + 750–19 + 850 (Fig. 2). Slopes which suffered the most intense cracking during the earthquake were noticeably the sites of renewed displacement or shallow landsliding during the monsoon storm.

3. *Displacement and failure of retaining walls.* Both masonry and gabion gravity retaining walls, up to 7 m and 10 m high respectively, had been constructed to support the road formation in this area (Fig. 2). None of the gabion walls failed in the earthquake. Most experienced relatively minor movement and settlement (generally <200 mm), with slight to moderate consequential cracking of the road pavement. In contrast, two sections of masonry wall failed completely by rotation and most of the other existing masonry walls between Ch 18 + 800–19 + 400 were displaced outwards by 0.5 to 3.0 m (Fig. 6). Based on mapping of these displacements and cracks within the pavement and backfill, the inferred mechanism of wall collapse is shown in Figure 7. In general, the worst-affected walls were not the highest but were those with minimum toe embedment (<1.5 m) as deduced from site observations and as-built wall elevation drawings.

What the mapping established

The conceptual models were used as a framework for assessing the likelihood of renewed instability or continuing displacement of slopes and walls and for making engineering judgements for the design of permanent remedial works at different sections along the alignment.

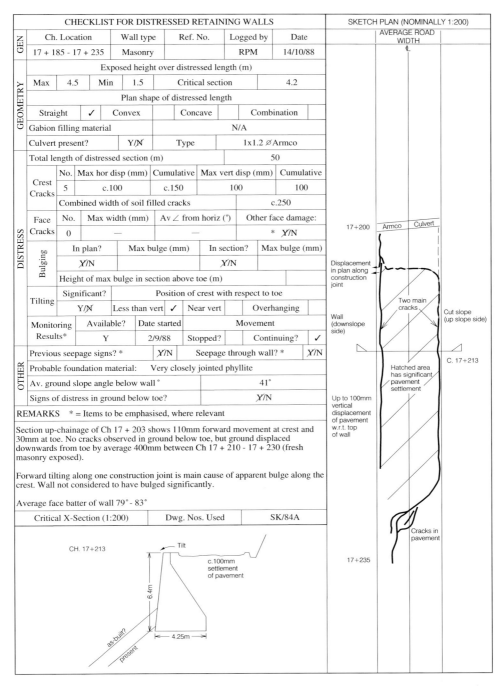

Fig. 4. Example of checklist used for distressed retaining walls.

Fig. 5. Two sections of roadline between Ch 20 + 100 and 20 + 200 destroyed by rockfalls from the quartzite pinnacles. These sections were rebuilt with new gabion retaining walls.

1. *Rockfalls and rockslides.* Almost all of the damaging rockfalls and rockslides were caused by the earthquake. The potential for renewed rockfalls or slides, or erosion of existing slide scars, was judged to be low. Hence the principles adopted for remedial works design were: (i) to rebuild the damaged sections of road with essentially the same cross-sections; and (ii) not to attempt to construct stabilization or improvement works to the existing rock slopes. (Full stabilization of the quartzite pinnacles was beyond the scope of reasonable improvement works and it was recommended that future seismically generated rockfalls should be an accepted risk.)

2. *Cracked and displaced soil slopes.* The crucial observation on these slopes was that cracking and the occurrence of debris slides appeared to be confined to the top 2–3 m of colluvial soil. Underlying fractured phyllites, where exposed, showed some evidence of loosening in the earthquake but there were no very large tension scars on the slopes or other signs of incipient deep-seated movement. The probability of a major hillside failure, deep enough to destroy both sections of road, was judged to be very low. Hence remedial works designs were based on: (i) excavating and removing cracked material to re-instate the cut slopes, or providing structural support in the form of gabion revetments wherever the angles of reinstated cut slopes would be greater than 50°; and (ii) grading over and sealing with compacted soil the larger cracks not removed by reinstated earthworks, to minimize infiltration in future storms.

3. *Displaced and failed retaining walls.* The response to the earthquake clearly demonstrated that the flexible gabion walls were superior to the masonry walls in their ability to accommodate differential foundation settlement and moderate seismic loads without serious distress

(A)

(B)

Fig. 6. The failed section of roadline (Ch 19 + 240–19 + 280) (**A**) shortly after the earthquake (October 1988) and (**B**) during construction of the remedial works (May 1989). In the foreground of (**B**) a new gabion retaining wall has been constructed to replace the failed section of masonry; behind this the displaced section of masonry wall has been repaired by foundation underpinning and buttressing.

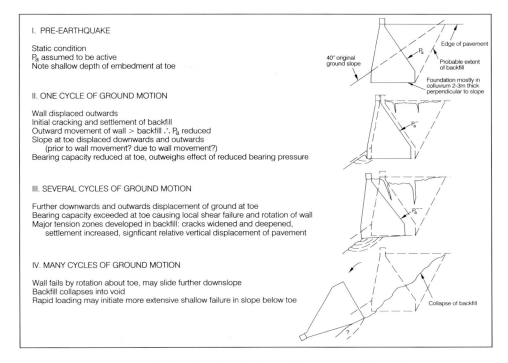

I. PRE-EARTHQUAKE

Static condition
P$_a$ assumed to be active
Note shallow depth of embedment at toe

II. ONE CYCLE OF GROUND MOTION

Wall displaced outwards
Initial cracking and settlement of backfill
Outward movement of wall > backfill ∴ P$_a$ reduced
Slope at toe displaced downwards and outwards
    (prior to wall movement? due to wall movement?)
Bearing capacity reduced at toe, outweighs effect of reduced bearing pressure

III. SEVERAL CYCLES OF GROUND MOTION

Further downwards and outwards displacement of ground at toe
Bearing capacity exceeded at toe causing local shear failure and rotation of wall
Major tension zones developed in backfill: cracks widened and deepened,
    settlement increased, significant relative vertical displacement of pavement

IV. MANY CYCLES OF GROUND MOTION

Wall fails by rotation about toe, may slide further downslope
Backfill collapses into void
Rapid loading may initiate more extensive shallow failure in slope below toe

Fig. 7. Inferred mechanism of collapse of masonry retaining walls.

(Martin 1982). Remedial works designs were based on the following guidelines: (i) the failed sections of masonry wall should be rebuilt in gabion with a minimum toe embedment of 2 m; (ii) distressed (rotated) masonry walls judged to be unserviceable should be removed and rebuilt in gabion; (iii) distressed masonry walls judged to be serviceable require substantial improvement to toe stability by foundation underpinning and/or buttressing; and (iv) distressed gabion walls generally require only partial backfill and pavement reconstruction.

The mapping and checklist results were of particular value in guiding the judgement on serviceability for items 3(ii) and 3(iii) above. Removal of displaced (but not completely rotated) pieces of masonry weighing up to several hundred tonnes was a major logistical constraint for remedial works design and construction. In the event, only three sections of severely distressed wall, of combined length 72 m, were judged completely unserviceable and required demolition, removal and replacement.

Construction of remedial works commenced in early 1989 and was substantially completed by mid-1990. The rebuilt and repaired walls have performed satisfactorily in the interim, with no undue further movements experienced. Further minor instability and erosion has occurred from time to time within the reinstated cut slopes, but this has been tolerated within the normal maintenance programme, as is the case for minor instability on other hill sections of the road.

Similar applications

Large-scale field mapping and use of checklists proved invaluable for appreciating the scale and nature of the damage in this study. The techniques were applied rapidly and cheaply, allowing completion of remedial works design drawings within eight weeks. This type of rapid post-construction assessment has widespread application to low-cost mountain roads subjected to earthquake or rainstorm damage.

The use of post-construction field mapping on the Dharan–Dhankuta Road follows a long tradition of engineering geomorphological and geological mapping applied at the reconnaissance, investigation and construction stages of the project (Brunsden *et al.* 1975; Fookes *et al.* 1985). This case study reinforces the view of Fookes *et al.* that field mapping in various forms is the fundamental tool for investigations of low-cost mountain roads.

Acknowledgements. The author was seconded to Messrs Roughton International to undertake the work described. The support given by Mr A. H. Cutler, Mr A. Murphy and Mr M. Pearce is gratefully acknowledged. This paper is published with the permission of the Department for International Development (U.K. Government), the Department of Roads of the Kingdom of Nepal and Messrs Roughton International.

References

BRUNSDEN, D., DOORNKAMP, J. C., FOOKES, P. G., JONES, D. K. C. & KELLY, J. M. H. 1975. Large scale geomorphological mapping and highway engineering design. *Quarterly Journal of Engineering Geology*, **8**, 227–253.

BRUNSDEN, D., JONES, D. K. C., MARTIN, R. P. & DOORNKAMP, J. C. 1981. The geomorphological character of part of the Low Himalaya of Eastern Nepal. *Zeitschrift fur Geomorphologie*, supplementband **37**, 25–72.

CROSS, W. K. 1982. Location and design of the Dharan–Dhankuta low cost road in eastern Nepal. *Proceedings of the Institution of Civil Engineers*, Part 1 **72**, 27–46.

DIKSHIT, A. M. 1991. Geological effects and intensity distribution of the Udayapur (Nepal) earthquake of August 20 1988. *Journal of the Nepal Geological Society*, **7** (special issue), 1–17.

FOOKES, P. G., SWEENEY, M., MANBY, C. N. D. & MARTIN, R. P. 1985. Geological and geotechnical engineering aspects of low-cost roads in mountainous terrain. *Engineering Geology*, **21**, 1–152.

MARTIN, R. P. 1982. Discussion on 'Retaining Walls and Basements'. *Proceedings of the Seventh South-east Asian Geotechnical Conference*, Hong Kong, **2**, 279–280.

ROUGHTON & PARTNERS. 1988. *Nepal Roads Remedial Works Project: Dharan–Dhankuta Road 1988 Damage Report.* Roughton & Partners International, Southampton, 2 vols (unpublished).

Terrain evaluation for military purposes: examples from the Balkans

C. P. Nathanail

School of Chemical Environmental Engineering, Nottingham University, Nottingham, UK

Aim/purpose

Terrain has influenced military commanders and the outcome of military operations since ancient times (Rose & Nathanail 2000). The aim of terrain evaluation in support of military operations revolves around gaining maximum operational advantage from the ground. In war this includes maximizing the mobility of your own forces and ensuring the survival of your troops and in turn denying both of these to the enemy's forces. Armed forces are also increasingly being used in peace keeping or peace enforcing roles. In such cases the aim of terrain evaluation is to support the humanitarian efforts of the military.

The United Nations Protection Force (UNPROFOR) was tasked with delivering humanitarian relief in Bosnia Herzegovina (Fig. 1). British troops were based in four main compounds at Vitez (School and Garage), Tuzla, Gornji Vakuf and Tomislavgrad. Terrain evaluation principles were used to advise on the feasibility of constructing water supply boreholes within the perimeter wire of each compound to ensure a secure supply of water independent of any of the warring factions.

Later, the NATO Implementation Force (IFOR) was tasked with ensuring that the provisions of the Dayton Peace Agreement were fully implemented (Haynes et al. 1997). One of the concerns was to ensure that the boundary between warring factions, which had been agreed on maps, was securely and accurately delineated on the ground. Terrain evaluation principles were applied to advise on the ease of installation and, malicious, extraction of border marker posts.

Terrain evaluation in Bosnia Herzegovina

UN and NATO forces have operated in Bosnia Herzegovina since the early 1990s in a variety of humanitarian relief and peace keeping roles. The terrain, climate and their interaction were major obstacles that required considerable engineering and logistics efforts to overcome.

Geological information was available for much of the country. However a lot of it was old, at a small scale or written in Serbian. Some of the geological maps had been produced by Yugoslav geologists for the Waffen SS during World War II. Anecdotal information indicated that the local geologists had deliberately introduced errors into those maps. This was subsequently confirmed during the reconnaissance of sites near Vitez in central Bosnia (Nathanail in prep).

Techniques used

The principal techniques involved in providing advice were: literature review, vehicle based reconnaissance, very limited (due to the security situation) walkover surveys, topographic and geological map appreciation, interviews with local experts and limited interpretation of satellite imagery.

Fig.1. Position of Bosnia Herzegovina (after http://www. graphicmaps.com/graphic_maps.htm).

From: GRIFFITHS, J. S. (ed.) *Land Surface Evaluation for Engineering Practice.* Geological Society, London, Engineering Geology Special Publications, **18**, 205–208. 0267-9914/01/$15.00 © The Geological Society of London 2001.

The literature review focused on obtaining geological maps of Bosnia and relevant papers on specific locations. Although publications in English were preferred, Serbian and German references were also found to be relevant. Although limited translation support was available, a good dictionary used by a terrain evaluation specialist was found to be more efficient in interpreting the technical terminology used in say borehole logs or illustrations in papers.

Site description: geology of Bosnia Herzegovina

Bosnia Herzegovina forms part of the Dinarids, in the Alpine mountain chain that marks the collision between the African and European continental plates. The Dinaric Alps comprise mainly carbonate rocks which have been folded parallel to the coastline of former Yugoslavia and subsequently uplifted and eroded. The former Yugoslavia contains many of the type localities of karst features. Many geomorphological terms have been derived from the Slovenian and Serbo–Croatian languages such as doline; polje – a linear depression; ponor – a sink-hole; uvala – a large depression due to the coalescence of dolines; and jama.

Tectonic and Palaeogeographic Zones

The Dinarids, named after Mount Dinara, comprise eight tectonic and palaeogeographic zones (Fig. 2) (Ager 1980).

The Dalmatian zone extends from a peninsula of Istria, on the Italian border, all the way down to Albania. The general northwesterly strike of the rocks is evidenced by the extremely elongated islands off the Croatian coast at Zadar and north of Dubrovnik (Fig. 2). The zone is dominated lithologically by limestone, other carbonates, and some Triassic evaporites.

The High Karst zone comprises white limestone mountains which have been eroded into deep valleys and caves. Much of the surface water has disappeared underground and given rise to the terrain known as karst. The limestones are in places intercollated with bauxites.

Flysch constitutes the main part of the narrow Bosnian zone between two major thrusts and represents a trough that persisted at least from the beginning of Jurassic until the late-Cretaceous (Ager 1980). Sarajevo is built on the Bosnian zone flysch. The Jurassic succession generally comprises pelagic limestone with chert passing up to radiolarites then into the thick Bosnian flysch. Much of this zone and the Serbian zone discussed below are concealed beneath post-orogenic Oligocene–Miocene molasse.

The Serbian zone largely comprises ophiolites whose dark colour give Montenegro its name. In places the pre-ophiolite continental crust comprising Palaeozoic basement is exposed overlain by early Triassic red sandstone. Within Bosnia the ophiolite sequence is overlain conformably by flysch which spans the Jurassic–Cretaceous boundary and corresponds with the flysch of the Bosnian zone to the SW. Ager (1980) reports that this is one of the rare places in Europe where the direct relationship of ophiolites and flysch can be seen.

Dearman et al. (1989) recognized that Bosnia lies within one of the tectonically active regions of the world where tectonism would affect the distribution of the principal types of weathering. The country is characterized by complex mountain relief and extensive geodynamic processes. The lower and middle levels suffer chemical weathering of the rocks resulting in the deposition of a thick clayey cover over bedrock. Abundant rainfall and groundwater enhances the development of landslides.

In the Dinarids, the carbonate series of the Dinarid zone and the inner Dinarian syncline region are subject to extensive karstification. Seismic activity is high with the southeastern end of the region being the most active. As the intra-montaine basin and some poljes lagged behind during the general tectonic uplift of the Dinarids in the Neogene, sediment that once formed part of a larger mantel are preserved in them (e.g. the basin of Sarajevo). The role of surface runoff in the evolution and dissection of the relief is evident; the flysch and molasse hills in eastern Serbia are less dissected than are the western parts of Slovenia which receive more rain.

Sources of information

A desk study to examine geological maps, papers and reports was carried out. This allowed the general

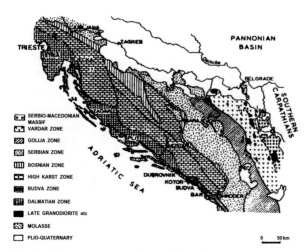

SERBIO-MACEDONIAN MASSIF
VARDAR ZONE
GOLIJA ZONE
SERBIAN ZONE
BOSNIAN ZONE
HIGH KARST ZONE
BUDVA ZONE
DALMATIAN ZONE
LATE GRANODIORITE etc
MOLASSE
PLIO-QUATERNARY

0   50 km

Fig. 2. Main tectonic zones in the Dinarides (after Ager 1980).

terrain and geological framework of the country and specific locations of interest to be determined – at least in outline.

Limited fieldwork was carried out at each site (Nathanail 1996, 1998). This amounted to a walkover survey within the perimeter fence of the UN compound. In one location (Vitex School) a hand auger was used to examine soils down to 2 m. In another, Tomislavgrad, a small borrow pit some 2–3 km away from the compound was examined to provide some indication of the nature of the soils in that particular valley.

Attempts to access locally held information proved variably successful. The offices of a quarry near the Vitex School compound were visited however the site appeared to be in use as a HQ for one of the warring factions and no information was retrieved. A former employee of the Local Authority in one area had logs of two public water supply boreholes some hundreds of metres from the Vitez Garage compound. Hand coloured maps of near surface materials, compiled principally for groundwater and aggregate resource assessment purposes, were made available, but only for inspection, at the Tuzla offices of a former public body.

The conceptual model – groundwater resources

At each of the five UN bases visited, a simple conceptual model of the near surface geology and hydrogeology was developed from the desk study information, terrain appreciation and walkover survey (Nathanail 1998). In each case the model focused on where groundwater resources were likely to be present in sufficient quantities and on potential threats to the quality of such water.

The results were used to support the decision to deploy the British Army's well drilling team for a number of successful tours of duty in the Balkans (Wye 1994).

The conceptual model – border marker posts

A map of surface soils was available. The requirement was to contribute to the selection of boundary marker posts to be installed along the agreed line between the various factions. The requirements were that posts be easy to install but difficult for belligerents to remove. The map legend was interpreted in terms of ease of installation (e.g. loose sand, soft clay versus bedrock or stiff clay) and ease of extraction (essentially the inverse of installation). The optimum ground being one in which it was moderately difficult to install but difficult to extract marker posts. Of course, the course of the boundary, being negotiated between the warring fractions, was not going to be affected by the ease or otherwise of marking it.

The Dayton Peace Agreement was initialled at Dayton, Ohio on 21 November 1995, following some six weeks of negotiations. To support the talks, the US military mapping agency – the Defence Mapping Agency (or DMA) – deployed a team of mapping specialists and equipment to Dayton. This team of approximately 40 technicians literally worked around the clock for those six weeks, producing 'the map' and accompanying overlays that were to define the partition of Bosnia Herzegovina. A digital terrain elevation model of the region onto which aerial photography of the whole of Bosnia was superimposed was created.

The operator then superimposed vector line data over this image and in particular the Inter–Entity Boundary Line between the former warring factions, or the IEBL as it became known. The computer image appeared on one screen and a co-ordinated map display on the other. Wherever the operator flew and drew a line on the aerial photography, it would automatically appear on the map.

In the end the two sides even disagreed on the type of marker they wanted. The Croat/ Muslim Federation wanted a ground mark for administrative purposes only, as they saw the IEBL somewhat akin to a British county boundary. The Bosnian Serbs, however, seemed to be using the former Berlin Wall as their model boundary. Finally they ended up with a marker similar to those used in UK to mark gas and water pipelines – a pole with a brightly coloured apex hat. Not surprisingly, this proved to be a collectors item for those who did not want any form of boundary.

What did the techniques reveal?

Despite the limited time and information available and restricted access to the ground, the use of terrain evaluation principles enabled a recommendation that a well drilling team be deployed from the UK to construct abstractions wells at each of the UN bases where British Troops were stationed. The team successfully constructed wells at each of the bases and then went on to construct wells in many other bases for other armed forces of other nations.

Terrain advice was also provided to assist in the planning for the eventual marking of the negotiated boundaries. However, in this instance terrain issues were secondary in the overall decision making process.

Similar situations and logical applications

The techniques of rapid terrain evaluation on the basis of insufficient information can be used in other military situations, in support of humanitarian relief operations, post disaster rehabilitation (Nathanail & Nathanail 1998) and in preliminary conceptual design of most major projects. Despite the incomplete database, the benefits of early appreciation of the terrain outweigh the relatively minor costs of staff time, information acquisition and travel related expenses to visit the site(s).

References

AGER, D. V. 1980. *The Geology of Europe*. McGraw Hill, London.

DEARMAN, W. R., SERGEV, E. M. & SHIBAKOVA, V. S. 1989. *Engineering Geology of the Earth*. Nauka Publishers, Moscow.

HAYNES, S., FAGG, A. & RIGBY, N. 1997. *Mapping For Peace – The Challenges of 250 Years Of Crisis Support*, presentation at Nottingham Trent University.

NATHANAIL, C. P. 1996. Environmental constraints on United Nations operations in Bosnia Hercegovina. *In*: COULSON, M. G. & BALDWIN, H. (eds) *Proceedings of the International Symposium on the Environment and Defence*, NATO CCMS Report Number **211**, North Atlantic Treaty Organisation, Brussels, 257–263.

NATHANAIL, C. P. 1998. Hydrogeological Assessments Of United Nations Bases in Bosnia Hercegovina. *In*: UNDERWOOD, J. & GUTH, P. (eds). *Military Geology in War and Peace*, Geological Society of America Reviews in Engineering GEOLOGY, **XIII**, 211–215.

NATHANAIL, C. P. & NATHANAIL, J. F. 1998. Mitigating geohazards affecting mountain roads in northeast Somaliland. *In*: MAUND, J. G. & EDDLESTON, M. (eds) *Geohazards in Engineering Geology*. Geological Society, London, Engineering Geology Special Publications, **15**, 231–237.

ROSE, E. P. F. & NATHANAIL, C. P. 2000. *Geology and War*. Geological Society, London.

WYE, T. 1994. Well drilling in Bosnia: *The Royal Engineers Journal*, **108**, 149–153.

Hazard assessment in Eastern Taiwan

D. N. Petley

Department of Geography, University of Durham, Durham, UK

Introduction

The topography of Taiwan is dominated by the Central Mountain Range, which has steep, densely vegetated slopes, deeply incised valleys and fractured, unstable rock masses. Although most of the population is located on the low-lying coastal plains, there is a need in this rapidly developing island to construct transportation links through the mountains. Unfortunately, the steep topography makes development difficult and there have been many instances in which poor terrain evaluation has led to inappropriate engineering solutions.

This paper outlines how landslide hazard assessment is being used as part of the planning process for the development of infrastructure along one such road through the mountains.

The physical setting of Taiwan

Taiwan is located on the eastern margin of the Eurasian continental plate (Fig. 1). The island, which is approximately 385 km in length and 143 km in width, has formed as a result of the ongoing collision of the Philippine Sea (oceanic) and Eurasian (continental) plates. Damaging, shallow seismic activity is frequent. The most seismically active region is in the central portion of the east coast of the island. However, the recent 'Chi-Chi' earthquake of 21 September 1999 occurred in the centre of Taiwan and caused the loss of over 2100 lives and induced damage island-wide.

Terrain evaluation in Taroko National Park

Taroko National Park is located on the eastern margin of Taiwan 121°40′ E and 24°0′ N (Fig. 1). It is centred upon the drainage basin of the Li-Wu River, with terrain that varies from lowland plains to steep, rugged, densely vegetated mountainsides. Morphologically, the drainage basin is dominated by Taroko Gorge, a spectacular canyon up to 1000 m deep and 17 km long. The geology, which is highly complex, comprises marble, green schist, black schist, pelite, sandstone, shale, and isolated bodies of gneiss. The general structure is that of a complex fold and thrust belt formed as a result of east–west compression, but with a number of large faults and thrusts displacing and off-setting the units (Petley et al. 1997).

Running east–west through the park is the only substantial road across the Central Mountain Range, the Central Cross Island Highway (CCIH). The road runs along the base of Taroko Gorge, positioned between 15 and 50 m above the river-bed on a series of interconnecting ledges. Thereafter it is threaded along the valley sides of the Li-Wu River, gaining altitude towards the west in order to climb over the high mountain passes. Improvements to the highway, which have included road widening and the construction of new sections, and the development of tourist infrastructure, have caused considerable disturbance to the geomorphic system (Petley 1998a, b).

The study reported here was undertaken as part of the planning process for Taroko National Park. In recent years, considerable investment has been made in upgrading the infrastructure. Unfortunately, the success of this construction has been quite variable, with several areas being adversely affected by gullies, landslides and

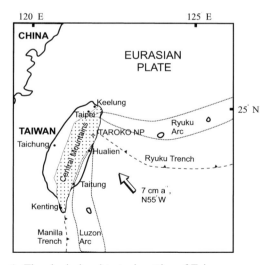

Fig. 1. The physical and tectonic setting of Taiwan.

From: GRIFFITHS, J. S. (ed.) Land Surface Evaluation for Engineering Practice. Geological Society, London, Engineering Geology Special Publications, **18**, 209–213. 0267-9914/01/$15.00 © The Geological Society of London 2001.

river channels. As a consequence, the National Park Headquarters has recently produced a comprehensive planning zonation to ensure that all new construction was undertaken in a manner that minimizes the hazard posed by natural processes.

The study reported here was undertaken for six centres of population along the CCIH. A draft development plan for each site had already been developed. The aim of this study was to utilize geomorphological terrain evaluation to assess the validity of the proposed scheme. Thus, the aims of this analysis were:

- to independently evaluate the geomorphology of the study areas;
- to establish the location of any potential geomorphological hazards within the planning area and to assess the likely impact on the site;
- to re-examine and interpret hazards that had been previously identified by National Park staff;
- to provide advice on the redesignation of areas in line with the hazard assessment where applicable.

Methodology

Each site was mapped using aerial photograph interpretation based on two epochs of 1:50 000 scale, vertical, colour, stereoscopic air photographs. Detailed geomorphological mapping was undertaken at each location during site visits. Where possible, sketches were made from a variety of oblique perspectives. Additional data were obtained on the nature of the local climatic system, and the frequency of occurrence of both typhoons, with their associated extreme precipitation events, and earthquakes. These data were supplemented by the acquisition of local information via an interpreter.

Thus, the approach adopted was similar to that used by Brunsden et al. (1975): namely, to construct a morphological map based on observed landforms, and then to use this as a base for the interpretation of morphogenesis and contemporary processes.

No attempt was made to undertake a risk assessment as the vulnerability of different planning elements was not considered (e.g. Dearman 1991; Brabb 1984). However, the study did give an indication of areas in which natural hazards have previously occurred and/or are likely to occur in the future, the nature of that hazard, its magnitude and frequency of occurrence. It provides the basic information that, combined with a vulnerability map, would allow a risk assessment map to be produced. In addition, it allows the National Park HQ to identify areas to which further resources should be directed.

Case study: Hualushi

Hualushi is a small farming community situated on a sharp bend in the highway, about 21 km along the road

to the west of Tien Hsiang and 1370 m above sea level. Structures at this site currently comprise:

- a large stall constructed without permission on the inside of the bend in the road, with a parking area in front;
- a number of other buildings on both sides of the road, some of which are used for trade and some of which are dwellings: many of the buildings have been built illegally;
- a terraced agricultural area in the northwest of the site, used for the cultivation of fruit trees.

The site is located on a narrow, gently sloping piece of land about 100 m above the Hualu Hsi, a tributary of the Wahel Hsi (Fig. 2). The river, situated to the east of the site, is deeply incised such that the slopes are steep, and they are densely forested. The slopes above the site to the west are also steep and densely vegetated, reaching a maximum height of 2253 m at Cheuh Shan, about 1.5 km away. This slope is traversed by the CCIH, and about 150 m above the site there is a hairpin bend in the highway. The slopes above Hualushi are characterized by a number of gullies, many of which flow onto the highway with no realistic means of controlling the water. In Hualushi itself, two large gullies cross the road, one on the apex of the bend in the centre of the settlement and one about 75 m uphill to the south.

The development plan for the site is as follows (Fig. 3):

- a narrow corridor has been reserved for further development of the highway;
- the stall and car park will be removed;
- small areas will be reserved for the development
- of administrative structures, tourist facilities and dwellings;
- the remainder of the area is to be protected and kept for conservation

In general the plan appears to be sympathetic to the environment. There is no evidence at the site of substantive mass movements or that seismic activity would trigger major problems. However, the failure to adequately consider the two large gully systems that cross the road in the vicinity of the township provides considerable concern. The gully that crosses the highway on the apex of the bend in the centre of the town provides the greatest cause for worry. This is a large feature (up to 5 m wide) draining a significant area of steep land above. Parts of this area have been developed for agriculture, whilst the remainder is forested. In 1997, flow in this gully resulting from the high precipitation intensities associated with the passage of Typhoon Amber overtopped the drainage culvert. Flow was diverted through the town, causing considerable damage to buildings.

Closer examination of the drainage basin for this gully casts light on the nature of the problem at this site.

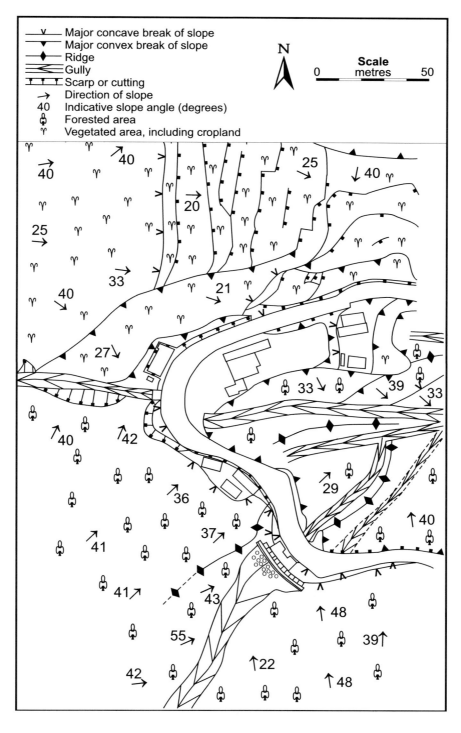

Fig. 2. Basic geomorphological map of Hualushi. Note the two major gully systems adjacent to the settlement.

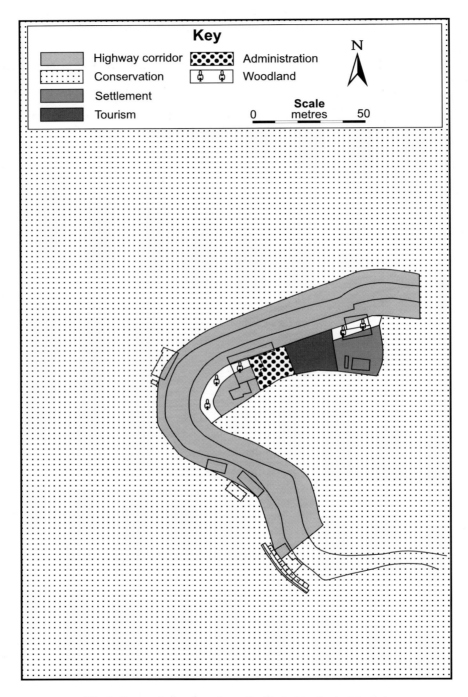

Fig. 3. Proposed planning scheme for Hualushi prior to this study.

(i) The drainage basin is relatively large, steep and densely vegetated, with about 700 m of relief. This provides the potential for short lag times, providing high discharges in typhoon events.

(ii) Some agricultural development of the drainage basin has occurred. The removal of the forest reduces the environment's ability to retain water, further reducing the lag time.

(iii) Above the township, the highway traverses the slope via hairpin bends. Examination of the road at this point suggests that inadequate provision has been made for allowing water in other gullies to cross the road, such that there is a tendency for water to utilize the highway as a fluvial channel. This has the potential to feed water from adjacent slopes into the gully in question, especially as one of the hairpin bends lies within the drainage basin. It is difficult to predict the path of water as the differential temporal activation of the gullies will lead to different fluvial pathways being created, but in some storms it is likely that extra water enters this gully via this route.

(iv) Upon examination of the culvert under the road, it was observed that it was clogged with rubbish. If the pipe is choked then all of the flow must be routed across the road and through the town, creating a significant hazard.

The other gully that provides some concern is the large feature 75 m or so up the road. Here another very large gully crosses the highway via a culvert. At this site the culvert is large and appears to be well constructed. However, there are signs that the road has suffered damage at this site, perhaps suggesting that water flows across the road at this location during high flow events. This is a source of concern as this water has a good chance of flowing down the road and through the settlement, possibly combining with the flow in the other gully. Similar concerns exist as at the previous gully in relation to the road above, although there is potential in this case for the road to remove water from the system. At present, this is difficult to predict and must not be relied upon. Fortunately, the catchment is not used for agriculture. However, this does not detract from the concerns about these features.

The gullies on the eastern side of the highway are deeply incised and active during storms. Care is also needed to ensure that these gullies are kept clear of development and garbage.

No major slope problems were identified at the site, although care is needed to ensure that the slopes adjacent to the incised gully systems are stable if used for devel-opment. Great care is needed to ensure that the slopes at this site are not disturbed by, for example, road building, which could trigger potentially serious problems.

Discussion and conclusions

The case study presented above demonstrates how terrain evaluation in the form of geomorphological mapping may contribute to the processes of creating development plans. The terrain evaluation demonstrated that the plan had limitations in that it failed to consider properly the hazards associated with the gully systems. As a result, a new plan has been put in place to allow for these features and thus to protect the infrastructure and inhabitants.

Use of similar methods to assess the hazards associated with cliffs above the road (Petley 1998*b*) has demonstrated that the accuracy of the method is high. Variations in the nature of the processes operating in the geomorphological hazards and in their frequency and magnitude are likely to occur, and these problems are likely to be exacerbated by variations in the techniques used to identify, examine and classify from one practitioner to another. Fortunately, the technique does not need to suffer from the problems associated with qualitative assessments of hazard rating. However, overall the use of geomorphological mapping for the evaluation of terrain during the planning process has considerable merit. As it is relatively low in cost and accurate, the system should be adopted widely.

References

BRABB, E. E. 1984. Innovative approaches to landslide hazard and risk mapping. *International Symposium on Landslides* (Toronto, Canada), **1**, 307–323.

BRUNSDEN, D., DOORNKAMP, J. C., FOOKES, P. G., JONES, D. K. C. & KELLY, J. M. H. 1975. Large scale geomorphological mapping and highway engineering design. *Quarterly Journal of Engineering Geology*, **8**, 227–253.

DEARMAN, W. R. 1991. *Engineering Geology Mapping*. Butterworth Heinemann, Oxford.

PETLEY, D. N. 1998*a*. Engineering hazards in the Taroko Gorge, Eastern Taiwan. *In*: MAUND, J. G. & EDDLESTON, M. (eds) *Geohazards and Engineering Geology*. Geological Society, London, Engineering Geology Special Publications, **15**, 125–132.

PETLEY, D. N. 1998*b*. Geomorphological mapping for natural hazard assessment in neotectonic terrains. *Geographical Journal*, **164**, 183–201.

PETLEY, D. N., LIU, C-N. & LIOU, Y-S. 1997. Geohazards in a Neotectonic Terrain, Taroko Gorge, eastern Taiwan. *Memoir of the Geological Society of China*, **40**, 135–154.

Slope instability within a residential area in Cleveland, UK

P. J. Phipps

Mott MacDonald, Croydon, Surrey, UK

Purpose of survey

An integrated geomorphological survey involving detailed geomorphological field mapping and associated desk studies is a powerful tool in identifying the locations and nature of slope instability within built-up areas. When supplemented with existing or specifically designed ground investigation, initial instability processes and mechanism models can be re-evaluated and refined. This paper provides an example of a geomorphological survey at Yarm, County Cleveland, which aimed to delineate unstable areas and address the potential causes of ground movements within a residential area.

The site

Valley Drive and Denevale are situated at the far eastern end of Yarm, on the south side of the River Tees, near its confluence with the River Leven. The natural valley side slopes are about 10–15° in this area and rise up towards a plateau surface at 30 m A.O.D., some 20–25 m above the Tees Valley. In order to facilitate development on the valley side slopes, extensive cut and fill operations have been carried out. Artificial slopes steepened up to 37° and vertical walls to 3 m high were also evident.

The published 1:10 560 scale British Geological Survey map (1976) and the 1:50 000 scale British Geological Survey map (1987) indicate Triassic Sherwood Standstone formation (formerly 'Bunter' Sandstone) as outcropping within the river channel. The valley side slopes have been developed in thick superficial deposits which comprise Upper and Lower Glacial Till units, separated by Glacial Sands and Gravels overlying Laminated Clay. These materials are believed to have been deposited during the last Cold Stage (Devensian 80 000–10 000 years BC) (Catt 1991).

The published geological maps show extensive areas of landsliding within both the Tees and Leven Valleys. Prior to the start of operation of the Tees Barrage in December 1994, a confidential survey of the landslides within the valley was carried out by an independent consultant in 1989 that provided a record of instability features.

Techniques used

Detailed geomorphological field mapping was undertaken within the Valley Drive/Denevale area of Yarm at an original scale of 1:500. The base maps were obtained from enlargements of existing 1:1250 Ordnance Survey mapping. A total area of 400 m × 150 m was mapped in a two-day period. The mapping methods were based on those described in Cooke & Doornkamp (1990) and Savigear (1965).

The field mapping was supplemented by a desk study which included a review of historical Ordnance Survey maps and available historical stereoscopic aerial photographs. These assisted in the identification of slope form, water and drainage features, and areas of instability to be set against the residential development and landscaping of the area. Some monitoring information was also available from existing inclinometers, piezometers and groundwater wells.

The conceptual model

The natural slopes on this section of the Tees Valley were extensively modified by cut and fill operations for housing development. At the outset of the geomorphological field mapping the importance of understanding the relationships between artificial slopes and the natural geomorphology to aid stability assessments were recognized. The geomorphological maps produced delineated the different slope forms and their associated instability features (Fig. 1). Three main types of slope instability were identified within the mapped area consisting of the following.

1. *Compound landslides* had progressed up-slope from the river edges. Failure occurred through *in situ* Glacial Till and had typically incorporated variable amounts of overlying fill. The landslides consisted of a series of failures that appeared to move along non-rotational failure surfaces at depth. Existing morphology was represented by a number of failed blocks and associated scarp slopes. Surface expression of the failed areas appeared to be controlled by the distribution of overlying weaker fill.

From: GRIFFITHS, J. S. (ed.) *Land Surface Evaluation for Engineering Practice*. Geological Society, London, Engineering Geology Special Publications, **18**, 215–219. 0267-9914/01/$15.00 © The Geological Society of London 2001.

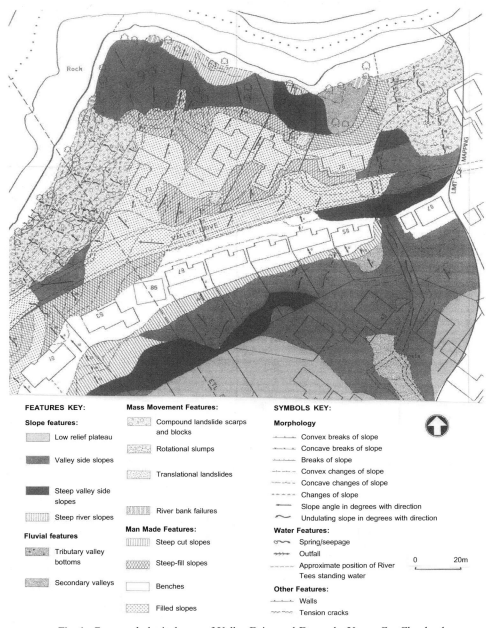

Fig. 1. Geomorphological map of Valley Drive and Denevale, Yarm, Co. Cleveland.

2. *Translational landslides* were typically limited to locally oversteepened cut and/or fill slopes. The failure surface was dominantly planar and sub-parallel to the ground surface. The depth of failure typically did not exceed 2 m. Surface form was often poorly defined on the ground in the study area.

3. *Rotational slumps* developed in fill slopes, which consisted of single rotational failures.

Each of these failure types had different implications regarding their risk to various structures and potential mitigation measures. The distribution of slope instability

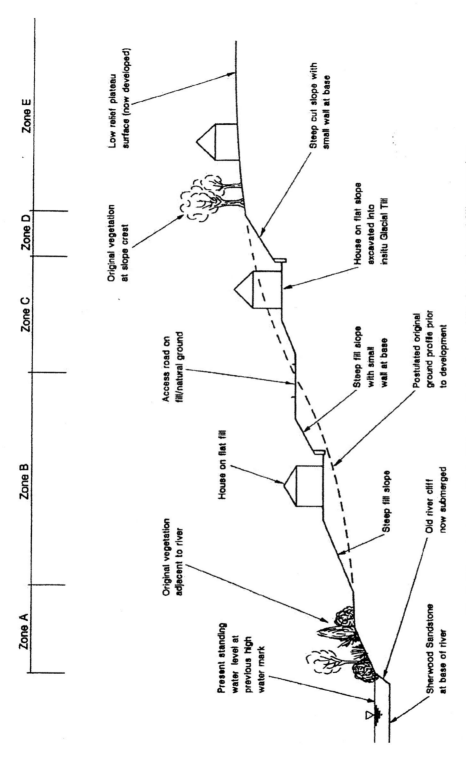

Fig. 2. Schematic slope profile identifying the effects on local stability from cutting and filling operations.

was found to be fundamentally controlled by the artificial modifications to the natural slopes. Five zones could be delineated within the side slopes with specific stability characteristics. These are presented in Figure 2 and described below.

Zone A. Slopes adjacent to the river that had not been altered by excavation activities. Vegetation was that existing prior to development. Fill slopes directly upslope within Zone B had the effect of loading the rear of these Zone A slopes.

Zone B. Extensive landscaping of fill slopes that had been built up on the existing lower valley side slopes. These activities generally steepened the overall slope angles and increased shear stresses in the underlying slopes, including those in Zone A.

Zone C. Slope angles have been decreased by excavations in the existing natural slopes, mainly for house benches.

Zone D. Slopes steepened by excavation. Small walls that may not have been designed as retaining structures were often at the base of such cut slopes.

Zone E. Original, more gently inclined upper valley side slopes and plateau surfaces that have been developed without the need for extensive cut and fill operations.

What the survey established

The geomorphological survey identified the distribution of three major landslide types. Both the translational landslides and rotational slumps were confined to individual cut/fill slopes and did not tend to exceed 3 m depth below ground level. Compound landslides were mainly present in the natural side slopes of Zone A directly adjacent to the River Tees, although they did locally impinge into the lower fill slopes of Zone B. Existing ground investigations indicated that the depth of the compound landslides was controlled by the presence of an eroded Sherwood Sandstone surface roughly at river bed level.

Derivative ground behaviour maps were developed for the site which incorporated both the form of slope instability and the rates/magnitudes of recorded movements. Such maps provide a factual description of different recorded impacts associated with the various ground movement types in each slope category and identify the susceptibility of the slope categories to instability.

The potential for the site to be subject to major valley side failures was also addressed as events of this magnitude could not be mitigated against without recourse to major ground investigations and subsequent remedial works. Studies of other glacial till slopes with similar

geotechnical characteristics suggested that most instability, whether shallow or more deeply seated, occurred on slopes in excess of 25° (i.e. Chandler 1984). Failure tended to be independent of height although this factor usually controls the failure mechanism. Slopes that were less than 4 m in height and affected by instability tended to be subject to translational landslides or rotational slumps, with major compound landslides confined to slope sections that exceed 4 m in height.

The overall natural slope angles at the site did not exceed 15° and the majority of existing natural landslides in the Tees Valley were associated with slopes exceeding 20°. Slope instability in the study area tended to be confined to failures involving artificial slopes, although some natural instability related to bank erosion by the River Tees was also evident.

The occurrences of slope instability identified during the survey were not attributable to a single causal factor. Different forms of instability were present in various slope zones which had their own stability characteristics. Some of the causal factors that had an impact on the slope stability included the natural slope development with associated artificial modification to slope geometry and groundwater conditions. These were dominated by the cut and fill operations and usage of the Tees Barrage. Vegetation changes on the valley slopes related to development activity were also identified as a contributory factor to local shallow instability.

Similar applications

The type of geomorphological survey adopted for the residentially developed valley side slopes at Yarm can be applied to a variety of geomorphological situations for different project requirements. However, the basic tenet, that the geomorphological field mapping records the existing landforms and processes and the desk study element provides general information on the site and its history, is a very powerful tool to assist land surface evaluation for engineering purposes.

For the Yarm site it was a fundamental requirement to delineate and classify the various instability types. Subsequent information enabled a factual description of the ground behaviour at the site to be produced which then formed the basis for addressing the risks to individual properties and the potential for major valley side slope failures.

References

BRITISH GEOLOGICAL SURVEY. 1976. *Sheet NZ 41 SW, 1:10 560.*
BRITISH GEOLOGICAL SURVEY. 1987. *Sheet 33, Stockton Solid and Drift, 1:50 000.*

CATT, J. A. 1991. The Quaternary History and Glacial Deposits of East Yorkshire. *In*: EHLERS, J., GIBBARD, P. L. & ROSE, J. (eds) *Glacial Deposits in Great Britain and Ireland*. Balkema, Rotterdam, 185–193.

CHANDLER, R. C. 1984. Recent European experience of landslides in over-consolidate clay and soft rocks. *Proceedings of the 4th International Symposium on Landslides*, Toronto, September 1984.

COOKE, R. U. & DOORNKAMP, J. C. 1990. *Geomorphology in Environmental Management*, second edition. Clarendon Press, Oxford.

SAVIGEAR, R. A. G. 1965. A technique of morphological mapping. *Annals of the Association of American Geographers*, **53**, 514–538.

Ground models for the design and construction of a high-speed rail link

P. J. Phipps

Mott MacDonald, Croydon, Surrey, UK

Purpose of survey

Engineering geomorphological, geological and geotechnical information, pertinent to a long linear engineering project, was required to provide support to a Parliamentary Bill and assist the future design and construction of the Channel Tunnel Rail Link. A wide variety of reports, maps and borehole logs had been collated for locations in the vicinity of the route alignment. These were to be supplemented with Phase I ground investigations for limited locations along the route during the winter of 1993/94. Phase I information was to be incorporated with the desk study information to provide an initial classification of the route geomorphology and ground conditions, and provide a sensible basis from which to design subsequent Phase II ground investigations.

The site

The Channel Tunnel Rail Link (CTRL) is to be constructed between St Pancras and the Channel Tunnel Railway Terminal at Cheriton, near Folkestone. The CTRL is a high-speed railway, 108 km in length, and its construction will incorporate a variety of engineered structures. The route traverses a wide range of topography, and solid and drift geology, which produce a wide variety of ground conditions to be considered for design and construction purposes.

At a macroscale the CTRL is located in three distinct geomorphological areas: the London Basin, the North Downs and the northern limb of the Weald. A wide variety of solid and drift geologies were evident from available British Geological Survey mapping: these included Lower London Tertiary deposits, Pleistocene Terrace deposits and recent alluvium in the London Basin, the Upper Cretaceous Chalk-dominated landscape from the Thames Valley to the North Downs, and the Cretaceous strata of the Weald.

The geomorphological and structural geological features of the route alignment had been developed mainly as a function of post-Cretaceous landscape evolution during the Tertiary controlled by tectonic and isoeustatic fluctuations. The landscape was further significantly modified by periglacial conditions during the

Pleistocene cold phases and to a lesser extent by more quiescent inter-glacial phases (Jones 1980). The most recent effects on the landscape have been from the geomorphological processes acting during the present Holocene, and from the impacts of human activities.

Techniques used

It was recognized early on at the Phase I stage that to provide a relevant structure for the ground investigations, associated studies and presentation of the information, then a robust and appropriate methodology had to be adopted. Terrain systems mapping was identified as providing a suitable hierarchical framework for producing initial ground models, designing the Phase II ground investigations, and incorporating engineering geological and geotechnical information back into the ground models. The actual techniques used were based on those described in Cooke & Doornkamp (1990) and for engineering projects in Brink et al. (1968). Satellite imagery interpretation at 1:50 000 scale was undertaken to assist with the delineation of the initial terrain systems.

Detailed geomorphological field mapping at 1:1250 was carried out for a number of type site areas to provide detailed information on specific landscape characteristics including morphology, near-surface materials, slope instability, drainage and areas of soft ground (i.e. Brunsden et al. 1975). The detailed field mapping was also supplemented by aerial photograph interpretation for the entire route using stereoscopic colour photographs at 1:2500 scale.

The conceptual model

The wide variety of ground conditions expected over the route alignment and the various engineering structures that had to be designed and constructed for those ground conditions required a robust and concise way of recording, assessing and generating the relevant information (Waller & Phipps 1996). The terrain evaluation had to provide a basis for designing relevant and cost-effective Phase II ground investigation and be flexible enough to allow revision by it.

From: GRIFFITHS, J. S. (ed.) Land Surface Evaluation for Engineering Practice. Geological Society, London, Engineering Geology Special Publications, 18, 221–225. 0267-9914/01/$15.00 © The Geological Society of London 2001.

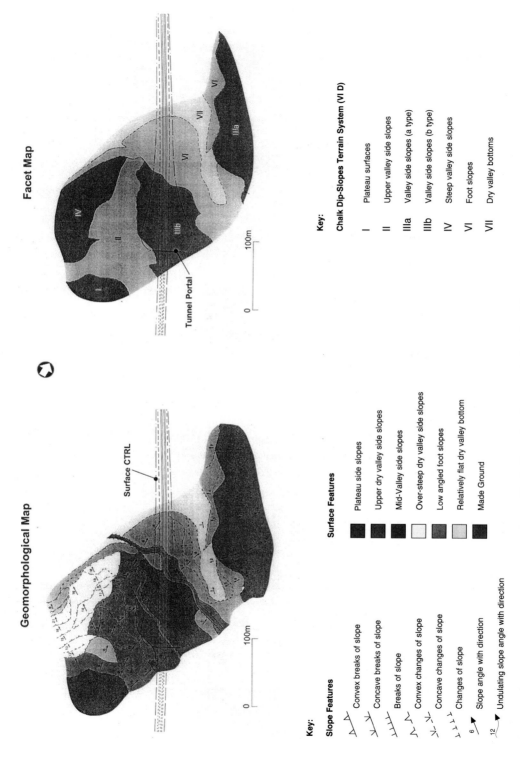

Fig. 1. Examples of geomorphological and facet maps.

Initial interpretations of satellite imagery were referenced against available published geological mapping to provide small-scale division of the route and its environs into terrain systems. These divisions reflected characteristic topography, drainage and vegetation correlated to geomorphology and geology.

The detailed geomorphological field mapping was undertaken for type site areas within those terrain systems where it was expected to have benefits for the terrain evaluation and ground model development (Fig.1). No mapping was therefore undertaken in the floodplain areas or the London Basin where the CTRL would be in a deep tunnel. Two to three days were typically taken to map areas up to $1\,km^2$ that translated to over $20\,km$ of the CTRL route length.

Following the detailed geomorphological mapping, the type site areas were divided into typically less than eight terrain facets. The facets summarized the detailed field data as each one comprised characteristic slopes, drainage, weathering profiles, associated relict and contemporary processes, and engineering concerns. Ground profile information was also summarized for each facet from assessment of Phase I and historical borehole data, field observations of drift and solid geotechnical exposures, and hand augering. These data were presented as a series of summary tables accompanying the facet maps

Table 1. *Example extract of a facet summary table: Terrain system XI: Hythe Beds surface*

Facet code	Facet name	Slopes	Materials	Drainage	Contemporary processes	Relict processes	Engineering concerns
I	Hythe Beds plateeau surface	0–3°	Typically 0–2 m of sandy clays and silty sands with fragments of flint chert and ragstone (head) overlying 18 m or less of interbedded sands, sandstone and limestone. The basal members consist of silty or clayey sands and thin to medium bedded weak to moderately strong sandstones (Hythe Beds). Sands at the base grade into silts and clays in the transition zone with the underlying Atherfield Clay.	No evidence of ephemeral drainage in the study area. Plateau surface well drained. Slopes are oversteepened to 3° around the heads of spring sapped valleys (see III).	Very slow movement of superficials downslope by creep and hillwash processes. Slow solution weathering of underlying Hythe Beds possible.	Solution weathering from Tertiary of upper layers. Periglacial processes during the last cold phase have led to reworked materials to 3 m depth or more, and fracturing of the underlying Hythe Beds.	Variability of depth to rockhead. Variability in extent and type of superficial deposits. Presence of palaeoshear surfaces in superficial deposits. Presence of infilled quarries with little or no surface expression.
II	Cambered Hythe Beds zone	3–6°	See I. Nature of the cambering leads to a greater thickness of superficials masking the underlying Hythe Beds. Up to 4 m of head recorded, typically consisting of stiff orange brown and green clays with a little angular fine to coarse flint or quartz gravel. Disturbed Hythe Beds caused by the cambering of the plateau surface near the scarp front leads to the formation of gulls and open joints, which tend to fill with superficials.	No perennial drainage features. Slopes are oversteepened to 8° around the heads of spring sapped valleys.	Slow processes of hillwash and creep. Cambering of blocks of Hythe Beds downslope may be occurring at a very slow rate. Spring sapping at Atherfield Clay/Hythe Beds interface.	Dislocation and cambering of Hythe Beds blocks especially during the Anglian cold phase. Evidence of former perennial drainage in dry valley forms.	See I. Lateral variability in Hythe Beds sandstone bands. Possible presence of large-scale palaeoshears developed during the Quaternary.

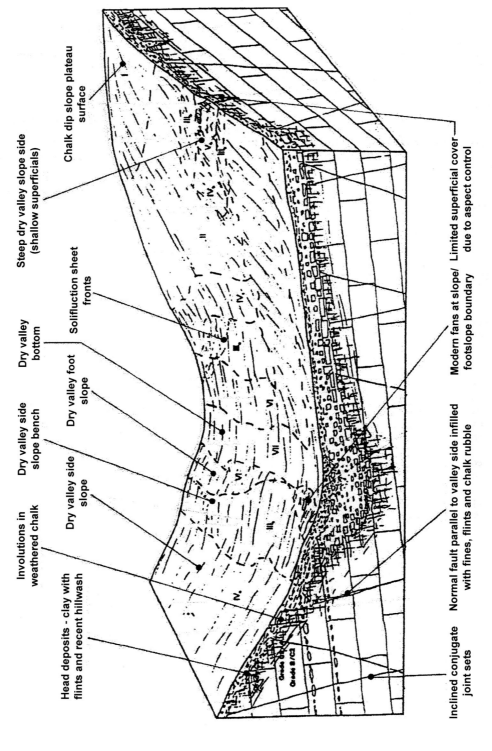

Chalk dip slope plateau surface

Steep dry valley slope side (shallow superficials)

Solifluction sheet fronts

Dry valley bottom

Dry valley side slope bench

Dry valley foot slope

Dry valley side slope

Involutions in weathered chalk

Head deposits - clay with flints and recent hillwash

Limited superficial cover due to aspect control

Modern fans at slope/ footslope boundary

Normal fault parallel to valley side infilled with fines, flints and chalk rubble

Inclined conjugate joint sets

Fig. 2. Example of a three-dimensional block diagram: Nashenden Valley.

(Table 1). Facets typically included dry valley bottoms, steep valley side slopes and plateau surfaces (Fig. 1).

Many of the boundaries between the terrain facets were morphological and the aerial photograph interpretation was used to extend the facet mapping away from the type site area in each terrain system. The process proved to be very rapid and accurate to within 5 m for areas of open fields. Ground truthing was undertaken to verify the facet boundaries.

What the survey established

A total of 14 terrain systems were identified through which the CTRL route was to pass. Within each of the terrain systems outside the London Basin tunnels area, facet mapping was produced at a scale of 1:2500 from a combination of detailed geomorphological field mapping and aerial photograph interpretation.

Relevant engineering geomorphological and geological information to assist in the design and construction of the CTRL was synthesized for each facet. Summary tables (Table 1) and three-dimensional conceptual ground models that showed the relationship between surface features, contemporary processes and the underlying geotechnical materials and structure were produced to aid in visualization of the issues (Fig. 2).

The majority of the Phase II site investigations were designed on the basis of the preliminary facet mapping. Locations of boreholes or trial pits were often decided by a requirement to improve the subsurface information for certain terrain facets. Both facet boundaries and terrain system boundaries were often subject to specific Phase II investigation, especially when the boundaries were unclear or contentious. In this way the terrain evaluation could be refined in response to the improved data and the knowledge of geomorphological, geological and engineering characteristics and issues within the facets. Anomalous features identified in the field and from the aerial photograph interpretation were also investigated in a further attempt to understand their origin.

Similar applications

Terrain evaluation for civil engineering purposes is rarely utilized, mainly owing to the lack of specific detailed information required by designers and contractors. However, such studies can provide appropriate summaries of engineering geomorphological features and typical ground conditions. They may also provide the basis for designing more detailed ground investigations to address specific issues or enable more subsurface information to be obtained for terrain systems or individual facets where appropriate information is lacking.

The technique of terrain evaluation can be adopted for any long linear engineering project including roads, railways and pipelines, and additionally for resource surveys. A large amount of data can be synthesized into a simpler format, or conversely limited data in certain locations can often be extended into similar terrain systems that are lacking in information.

References

BRINK, A. B. A., PARTRIDGE, T. C., WEBSTER, R. & WILLIAMS, A. A. B. 1968. Land classification and data storage for the engineering usage of natural materials. *Proceedings of the Symposium on Terrain Evaluation for Engineering*. Australian Road Research Board, 1624–1647.

BRUNSDEN, D., DOORNKAMP, J. C., FOOKES, P. G., JONES, D. K. C. & KELLY, J. M. H. 1975. Large scale geomorphological mapping for highway engineering. *Quarterly Journal of Engineering Geology*, **8**, 227–253.

COOKE, R. U. & DOORNKAMP, J. C. 1990. *Geomorphology in Environmental Management*. Clarendon Press, Oxford.

JONES, D. K. C. 1980. *The Shaping of Southern England*. Institute of British Geographers, Special Publication, No. **11**, Academic Press.

WALLER, A. M. & PHIPPS, P. J. 1996. Terrain systems mapping and geomorphological studies for the Channel Tunnel Rail Link. *In*: CRAIG, C. (ed.) *Advances in Site Investigation Practice*. Thomas Telford, London, 25–38.

Slope stability hazard assessment: Coalport Railway Bridge, Shropshire

D. T. Shilston

W. S. Atkins Consultants Ltd, Epsom, Surrey, UK

Objectives

Escalating costs, disputes and delays can arise during construction as a result of inadequate research at the early stages of project development. The assessment of slope stability hazard described in this case history illustrates the relative ease, and therefore low cost, of rapidly accessing and evaluating a wide range of desk study and walk-over survey data pertinent to a construction project.

The project

Coalport is a village on the north bank of the River Severn in the Ironbridge Gorge, near the new town of Telford, Shropshire. Coalport Railway Bridge (BC 17) is at the site of the former Coalport Railway Station. It is immediately adjacent to the much older Coalport (river) Bridge and carries the road from the river bridge across the former railway line (see Figs 1 and 2).

As part of a programme of bridge inspection and strengthening, the Commission for New Towns engaged WS Atkins to assess the condition of Railway Bridge BC 17 and make recommendations for remedial measures.

An initial inspection of the Railway Bridge concluded that it was in a poor state of repair and that it may be at risk from ground movements caused by landslipping, being in an area that is well known for the presence of landslips (Carson & Fisher 1991). A more detailed desk study, including a walk-over survey, was therefore carried out specifically to assess the landslide hazard at the railway bridge and to help design the ground investigation that was planned as part of the design of remedial works to the bridge's abutments.

The natural topography of the site comprises the bank of the River Severn leading up to a narrow river terrace and then the steep sides of the gorge whose crest is about 55 m above the river terrace. In the vicinity of the Railway Bridge, the natural topography has been extensively modified by the construction of Coalport Bridge (built in 1818 on the foundations of an earlier timber bridge of 1780; see Fig. 2); the Coalport Canal (1792); the Coalport Branch Railway with Railway Bridge BC 17 and railway station (1861); roads; various private houses; and recently a trunk sewer from Telford new town. The railway was dismantled in 1960, but the roads, bridges and trunk sewer are in everyday use.

The solid (bedrock) strata in this part of the Ironbridge Gorge comprise mudstones and siltstones with persistent sandstone layers of Carboniferous age (Fig. 3). They are mapped by the British Geological Survey (BGS) as dipping at a shallow angle to the east. Two north–south trending faults are mapped by BGS in the area, one of which (if extrapolated) would pass very close to the railway bridge. These strata are overlain by thin impersistant drift deposits and made ground. The BGS map shows large landslides adjacent to the Railway Bridge, but none at the bridge itself. However, man-made topographical changes have obscured much of the subtle geomorphological evidence for the original form of the hillside (and hence landslips) in the vicinity of the Railway Bridge.

Techniques

One key aspect of the studies for Railway Bridge BC17 was the integration of conventional engineering geological and geomorphological data (including mapping) with information about the site's history and human influence on the landscape. Only when this was done was it possible to arrive at a broadly based understanding of the landslide hazard at the site.

The classes of information used in the work are summarized on Table 1. They may appear to be unusually extensive but, in the author's experience, this is not the case. For many parts of Britain, particularly old industrial areas and areas of potentially adverse geotechnical conditions, investigators will find a large, varied and useful corpus of background and site-specific information (Shilston *et al.* 1998).

Three sources of information were found to be particularly useful:

- history of the Coalport branch Railway (Smith 1987);
- historical study of Coalport (river) Bridge (Trinder 1979);
- series of vertical stereographic aerial photographs of the site area: the earliest was from 1946 and shows the topography and geomorphology of parts now obscured by modern forestry.

From: GRIFFITHS, J. S. (ed.) *Land Surface Evaluation for Engineering Practice*. Geological Society, London, Engineering Geology Special Publications, **18**, 227–231. 0267-9914/01/$15.00 © The Geological Society of London 2001.

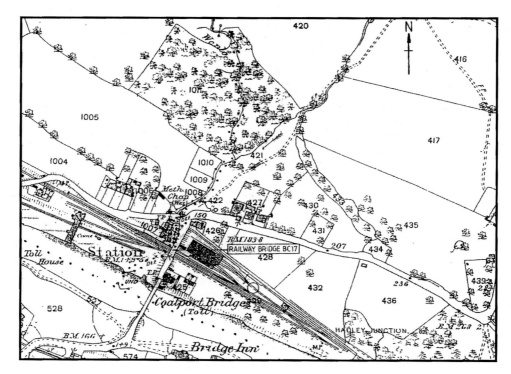

Fig. 1. Ordnance Survey map extract, 1883.

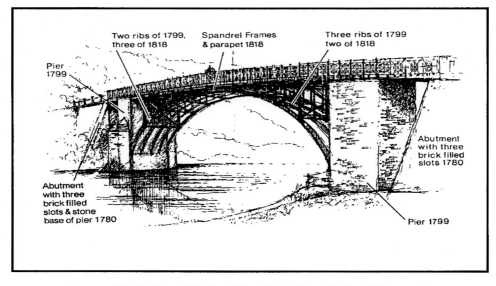

Fig. 2. View of Coalport River Bridge (from Trinder 1979).

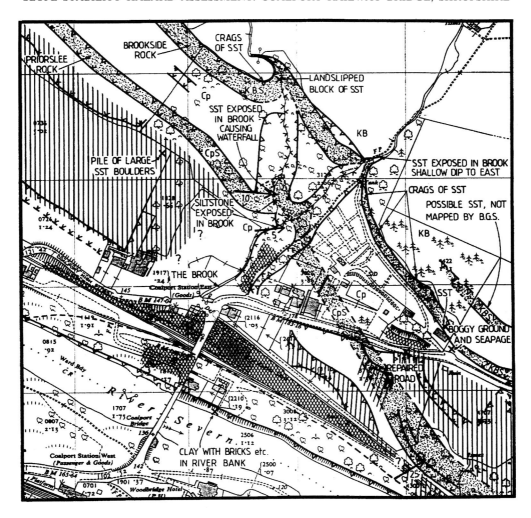

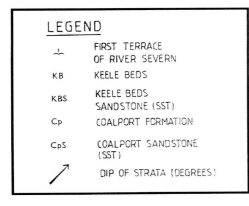

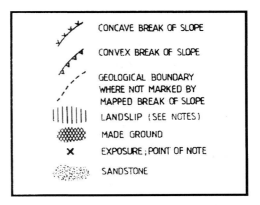

Fig. 3. Engineering geological sketch map.

Table 1. *Summary of information used in the desk study*

Class of information	Number of items
Topographic maps (historic and modern)	10
Aerial photographs (various dates)	6
Geological maps (including maps of boreholes, mines, wells and shafts	6
Technical papers and reports (site history)	9
Technical papers and reports (geology and geotechnical engineering)	14

The walk-over survey was carried out on two occasions. The first visit, which was at the start of the desk study, provided a general understanding of the site and its complex geomorphology and helped focus the desk study work on areas and technical aspects that seemed most likely to be significant. The second visit was made when the desk study was nearly complete. Its primary objectives were to provide 'ground truth' for the desk-based work and to investigate in the field specific areas of interest that had been identified. Both visits were made more cost-effective by being co-ordinated with visits to local sources of information.

The methodology for the engineering geological mapping and presentation of the sketch map (Fig. 3) follows that illustrated by Dearman & Fookes (1974). Figure 3 brings together information obtained by the desk study and walk-over survey: it illustrates the engineering geology, geomorphology and history of the site on a single figure. One or more synoptic maps such as this can be a very effective way of conveying complex information about the conditions and development of a site.

Results

The desk study work readily identified published information about landslips, elsewhere in the Ironbridge Gorge, that provided general guidance on slope instability in the area. Principal conclusions were:

- major deep-seated landslipping in the Ironbridge Gorge was initiated by rapid erosion of the gorge by the River Severn;
- shallower landslipping is present, both as a veneer to the deeper-seated landslips and in areas apparently lacking such landslips;
- some landslips are active today or have been stabilized by modern civil engineering works;
- recent reactivation of dormant shallow and deeper-seated landslips or accelerated movement of active landslips is generally the result of interference by man.

Building upon this body of background information, the desk study and associated walk-over survey enabled specific conclusions to be drawn about the characteristics of slope instability that could effect the Railway Bridge:

- very shallow, near-surface landslipping is likely wherever clay or mudstone strata are present on ground steeper than say 8°; that is, throughout the valley side;
- groundwater egress at springs is causing localized very shallow landslipping at the present day;
- apparently shallow landslipping is present to the west of the Railway Bridge on the broad expanse of mudstone outcrop which forms the lower zone of the hillside below a prominent sandstone unit;
- apparently complex landslipping may be present to the east of the Railway Bridge where the outcrops of two sandstone units drop in elevation down the valley side towards the River Severn;
- there is no firm geological evidence for landslipping in the area immediately uphill of the Railway Bridge;
- there is no geological evidence for instability affecting the River Terrace Deposits that occur at the base of the slopes adjacent to the River Severn;
- the satisfactory long-term (175 years) performance of the cast-iron Coalport (river) Bridge provides evidence of the stability of the lower part to the hillside as a whole; long-term squeezing or creep, as experienced for example some 2 km upstream at the Iron Bridge, does not appear to be taking place.

It was concluded that the known deeper-seated landslips in the vicinity of the Railway Bridge do not appear to be active at the present day, although there are signs of very shallow movements on most steep slopes and where there is groundwater egress. The known landslips can therefore be classified as dormant. However, the general experience of slope instability in the Ironbridge Gorge warns that movement could be initiated in the future by man-made or natural changes in the condition of the existing landslips, such as the ingress of water, placing of fill at their crests or excavation at their toes.

Evidence for landsliding was sought in the ground investigation for the remedial works at the Coalport Railway Bridge. The costs of doing so were much reduced by the findings of the desk study and walk-over survey which provided a rational basis for both the design and interpretation of the ground investigation.

Acknowledgements. The author would like to thank the Commission for New Towns for permission to publish this paper.

References

CARSON, A. M. & FISHER, J. 1991. Management of landslides within Shropshire. *In*: CHANDLER, R. J. (ed.) *Slope Stability Engineering*. Thomas Telford, London, 95–99.

DEARMAN, R. W. & FOOKES, P. G. 1974. Engineering geological mapping for civil engineering practice in the United Kingdom. *Quarterly Journal of Engineering Geology*, **7**, 223–256.

SHILSTON, D. T., HARRISON, E. N., PARSONS, A. S. & LEE, K. 1998. Giants' shoulders: the cost-effective use of geotechnical desk studies in civil and structural engineering. *In*: *The Value of Geotechnics in Construction*. Construction Research Communications Ltd, London, 25–36.

SMITH, W. H. 1987. The Coalport Branch. *British Railway Journal*, **19**, 399–415, 436–440.

TRINDER, B. S. 1979. Coalport Bridge: A study in historical interpretation. *Industrial Archaeology Review*, **3**, 153–157.

Evaluation in urban and industrial environments: the Docklands Light Railway, Lewisham Extension, London

D. T. Shilston, N. E. Harrison & D. J. French

W. S. Atkins Consultants Ltd, Epsom, Surrey, UK

Objectives

This case history describes the use of information obtained from the initial desk-based engineering geological and geotechnical studies carried out as part of the design of civil engineering works for the Lewisham Extension of the Docklands Light Railway (DLR) which was constructed in a heavily developed urban environment. The desk study aimed to make best use of available information in order to evaluate the terrain, limit the need for further ground investigation and provide geological and geotechnical information to the design teams as quickly as possible.

A desk study is an integral part of terrain and site evaluation and, wherever possible, should be carried out in conjunction with a reconnaissance survey (or walk-over) of the site. Together they are likely to be of greatest value and most cost-effective when carried out early in the investigations for a proposed development. They can provide the project team with an early indication of the conditions at the site, such information being essential for master planning and concept design. As the project is advanced the desk study findings become the basis for the planning and interpretation of the subsequent more expensive and lengthy stage of physical investigation by (for example) boreholes and trial pits.

Conceptually, it should be self-evident that it is cheaper to obtain and evaluate information that already exists than to obtain new investigation data. But this is not the only reason for carrying out desk studies. Their ability to recover information that cannot be obtained by other means is also very valuable. For example, in an urban or industrial environment, the study of historical maps, archive aerial photographs and other documentary information can provide vital insights into the past condition or uses of a site and the hazards that such use uses could pose to a new development (e.g. the presence of an ancient landslip, of old coal mines or of a potentially contaminating former industrial process). Thus, desk studies are part of the process of understanding and managing ground hazards and risks.

When planning desk studies it is helpful to use an aide-memoire or checklist of major classes of information that could be examined, such as that given in Table 1

Table 1. *Aide-memoire to classes of information for engineering geological, geomorphological and geotechnical desk studies*

Classes of information	Examples
Topography	Maps Aerial photographs Satellite imagery
Geology and hydrogeology	Maps Aerial photographs Satellite imagery Memoirs, reports, papers and books Mine and quarry records Thematic databases (e.g. in UK for landslides, natural cavities, etc.) Previous ground investigations
Environment and planning	Planning maps Aerial photographs Satellite imagery Archaeological site and historic building records Mine and quarry records Contaminated land records Landfill and waste disposal records Environmental statements Meteorological records River and coastal information
Site conditions, land use and history	Historical maps Historical documents Aerial photographs Satellite imagery Land use and planning maps
Site walk-over/reconnaissance survey	'Skilled eye' inspection of site and locality
Local knowledge and experience	Local history societies Library local studies departments Local press Neighbours Previous site users Construction records Building control office
Precedent	Case histories Construction records
Codes, standards and guidance	Professional bodies and institutes Government departments Research organizations and universities

From: GRIFFITHS, J. S. (ed.) *Land Surface Evaluation for Engineering Practice*. Geological Society, London, Engineering Geology Special Publications, **18**, 233–237. 0267-9914/01/$15.00 © The Geological Society of London 2001.

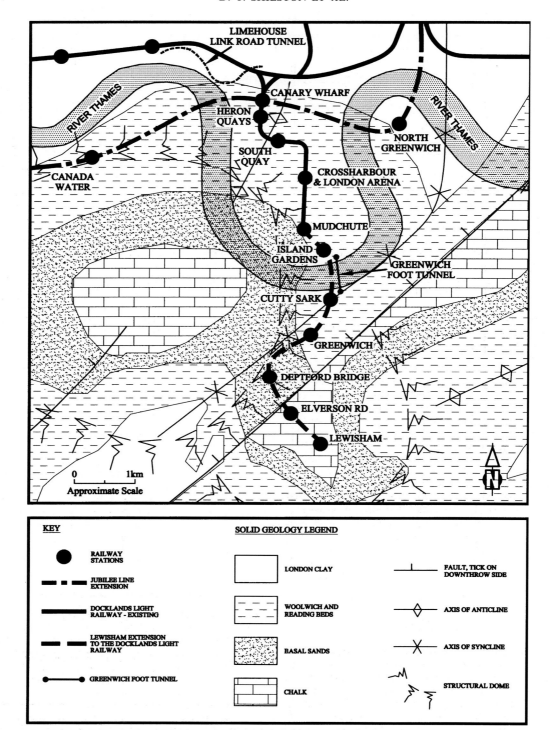

Fig. 1. Geological sketch map.

(Shilston *et al.* 1998). The table is based on the authors' own experience and can be augmented by guidance for individual countries or regions. As an example, for England, more detailed information on many of the sources mentioned is given in Transport Research Laboratory Report 192 (Perry & West 1996), *The Geologist's Directory* (Geological Society 2000) and the *English Local Studies Handbook* (Guy 1992).

The project

The Lewisham Extension to the Docklands Railway (DLR) runs approximately 4.2 km from Mudchute in London's Docklands, beneath the River Thames to Greenwich, and thence to Lewisham (Fig. 1). From north to south, the new railway comprises a cut and cover tunnel section from Mudchute to Island Gardens, twin bored tunnels under the River Thames to just west of Greenwich mainline station, a short length of cut and cover tunnel and then a surface section (partly on a viaduct) from Greenwich Station to Lewisham. As well as the railway line itself, two existing stations were replaced and five new stations constructed.

LRG Contractors, a joint venture between John Mowlem Civil Engineering plc and Mitsui-Nishimatsu, were awarded a contract to design and build the extension. WS Atkins were appointed by the Joint Venture partners to undertake the detailed design of the works.

This was a fast-track multidisciplinary project where information for preliminary design of the various structures was required almost immediately, to be updated with more accurate information as the design process progressed. The initial geotechnical tasks were to review the geotechnical parameters used in the tender design and to determine what additional site investigation would be necessary for detailed design to be carried out with confidence. Due to the very tight construction programme, the scope of any additional ground investigation had to be kept to a minimum and was restricted to a limited number of high quality boreholes and *in situ* tests. Collation and evaluation of existing information by means of a desk study were, therefore, central to the planning of the additional investigations and to the geotechnical design.

Techniques

Appropriate management, resourcing and presentation of the desk study work were critical to achieving the project requirements in the very short timescale available. A flexible team approach was adopted which could be tailored to the changing project priorities. Junior staff carried out most of the collection and preliminary review of reference information. For this to be successful, and in order that all relevant information was obtained, it

was essential that a senior member of staff comprehensively briefed team members on all relevant aspects of the project. In addition, experienced staff from the various disciplines involved in the project regularly reviewed the work as it progressed.

The desk study comprised a review of existing geological and hydrogeological data, site history records and construction precedent as summarized in Table 2. The table also summarizes the technical benefits of this work.

Presentation of information and results in an appropriate format for issue as formal reports can be time-consuming. An informal desk study file was maintained in the early stages of the project. It could be easily updated and held the key reference information along with summary tables similar to that reproduced in Table 2. In addition, figures that summarized information, such as overlays of historical maps, were produced so that the desk study findings could be properly assessed and presented to all parties.

Further details of the desk study are given in Shilston *et al.* (1998). Here we comment on one specific aspect of the work: the value of information about site history and construction precedent to the overall evaluation of the site. Such information can be particularly important where previous urban and industrial land uses are a significant component of the terrain through which a project is to be constructed.

Results

In common with many urban and industrial areas, London's Docklands and the London Boroughs of Greenwich and Lewisham have undergone numerous phases of development and redevelopment. Information on site history and construction precedent in the area was important in assessing the ground hazards faced by the project, determining the values of engineering properties for the design of the works and in the selection of construction methods. Examples of the use of, and conclusions drawn from, such information are given below.

Identification of construction hazards and risks

Groundwater Regime. The groundwater regime in this area of east London is complicated by the proximity of the Thames and rising groundwater levels in the underlying Chalk aquifer. This was of particular concern for the construction of the 23 m deep station box at Cutty Sark, which required the use of a very sophisticated dewatering system.

Unexploded bombs. The area was heavily bombed during the Second World War. There was therefore a possibility that unexploded bombs might be encountered during construction work.

Table 2. *Summary of availability and benefits of some of the information assessed during the desk study for the Lewisham Extension to the Docklands Light Railway*

Classes of information	Source	Description of information	Benefits
Geology and hydrogeology	British Geological Survey (BGS) maps and memoirs	BGS maps show the Greenwich Fault and associated folding and disturbances.	Better understanding of local geological structures, including faults, folds and unconformities and of potential hazards such as swallow holes.
	Borehole records from previous site investigations	BGS have numerous boreholes in the vicinity of the DLR extension.	Enhanced appreciation of ground conditions, thus limiting the amount of additional site investigation required.
	Published geological papers and literature	Howland (1991) provides a useful overview. Recent review by BGS has led to a reclassification of the Tertiary strata in the London area (Ellison *et al.* 1994).	As above
	Published hydrogeological papers and literature	CIRIA (1989) and Lucas & Robinson (1995) indicate that the DLR extension is in an area which may be affected by the rising groundwater table in the Deep Aquifer.	Better understanding of groundwater regime including influence of rising groundwater table.
	Mining records	Department of the Environment (1991) recorded mining activities in the Greenwich area.	Identified potential hazard of encountering shafts/deneholes.
Site conditions, land use and history	Historical maps	Miscellaneous historical maps from 1703 onwards and Ordnance Survey maps from 1869 to the present day.	Identified potential hazards associated with previous land use, including contaminated land, old foundations (piles, footings, etc.) and other construction works (river walls, basements, tunnels, etc.)
Precedent	Published case histories of construction projects in the area, particularly those involving deep excavations and tunnels.	Limehouse Link project experienced problems with dewatering part of the Tertiary strata which contained perched water and had high permeabilities. Records of the construction of the Greenwich Foot Tunnel and shafts (Copperthwaite 1901/02).	Obtain details of previous construction experience, particularly identification of potential construction problems and appropriate solutions.

Mining instability. Chalk and sand were mined in the nineteenth century in the Greenwich area.

Contaminated land. Ground and groundwater contamination were likely to be present from a wide range of former industrial activities at a number of localities along the route.

Building foundations. Piles and shallow foundations were likely to be present beneath some of the existing and demolished buildings in the vicinity of the surface sites and bored tunnels.

Identification of risks to adjacent structures

Ground movements caused by the works may affect some adjacent buildings. Where possible, construction records were obtained for prestigious or particularly sensitive structures. These included the existing Greenwich mainline railway station which was built in the nineteenth century and is a Grade II listed historic building. Drawings of the station's superstructure and foundations were acquired which proved invaluable in assessing the impact on the station of the proposed tunnelling and surface works.

Precedent experience

Greenwich Foot Tunnel. Opened in 1902, this tunnel connects Island Gardens to Greenwich and runs beneath the Thames close to the proposed route of the DLR's twin bored tunnels (see, Fig. 1). There is good documentation for the Foot Tunnel, including a contemporary paper published in the *Proceedings of the Institution of Civil Engineers* which describes the method of construction and the ground conditions encountered (Copperthwaite 1901/02).

Recent construction projects. The DLR Lewisham Extension is close to several major engineering projects such as the Limehouse Link Tunnel, Jubilee Line Extension and Canary Wharf (see Fig. 1). Information about these projects highlighted potential construction difficulties and particular ground hazards. These included, for example, difficulties encountered with dewatering the excavations for the Limehouse Link project (Cruikshank 1993).

Overview

Management of ground hazards and their consequent risks is an essential feature of civil engineering projects in complex urban and industrial terrain such as that in the area traversed by the Lewisham Extension of the DLR. The desk study work described in this paper played a fundamental role in identifying hazards and risks, developing ground models, and facilitating the development of effective mitigation measures through the selection of construction methods and the design of temporary and permanent works. These contributed to the successful construction of the project, the DLR Extension to Lewisham being opened in November 1999 some two months ahead of programme.

Acknowledgement. The authors would like to thank LRG Contractors for permission to publish this case study.

References

CIRIA. 1989. *The Engineering Implications of Rising Groundwater in the Deep Aquifer Beneath London.* Special Publication **69**. CIRIA, London.

COPPERTHWAITE, W. C. 1901/02. The Greenwich Footway Tunnel. *Proceedings of the Institution of Civil Engineers*, **150**.

CRUIKSHANK, J. 1993. Limehouse Link Supplement. *New Civil Engineer.* May.

DEPARTMENT OF THE ENVIRONMENT. 1991. *Review of Mining Instability in Great Britain.* Regional Report, South East England including Greater London, Vol. 1/ii.

ELLISON, R. A., KNOX, R. W. O'B., JOLLEY, D. W. & KING, C. 1994. A revision of the lithostratigraphical classification of Early Palaeogene strata of the London Basin and East Anglia. *Proceedings of the Geologists' Association*, **105**, 187–197.

GEOLOGICAL SOCIETY. 2000. *The Geologists' Directory.* Geological Society of London.

GUY, S. 1992. *English Local Studies Handbook.* University of Exeter Press.

HOWLAND, A. 1991. The Engineering Geology of the London Docklands. *Proceedings of the Institution of Civil Engineers. Part 1*, **90**, 1153–1178.

LUCAS, H. C. & ROBINSON, V. K. 1995. Modelling of rising groundwater levels in the Chalk aquifer of the London Basin. *Quarterly Journal of Engineering Geology*, **28**, Supplement 1.

PERRY, J. & WEST, G. 1996. *Sources of Information for Site Investigations in Britain* (Revision of TRL Report LR 403). Report 192, Transport Research Laboratory, Crowthorne.

SHILSTON, D. T., HARRISON, N. E., PARSONS, A. S. & LEE, K. 1998. Giants' shoulders: the cost-effective use of geotechnical desk studies in civil and structural engineering. *In*: *The Value of Geotechnics in Construction.* Construction Research Communications Ltd, London.

Section 4

Conclusions and Recommendations

Land surface evaluation: conclusions and recommendations

D. Brunsden & J. S. Griffiths

Emeritus Professor of Physical Geography, Kings' College, London, UK

Results from the Second Working Party Report

As discussed in the introduction to this report (**Griffiths & Edwards**), the Second Working Party on Land Surface Evaluation in Engineering Practice recognized that the subject area was so extensive that an 'instruction manual' or handbook, similar to previous working party reports, would not be appropriate. The end product, as presented, therefore, is a compilation of examples of techniques and case studies written by practitioners in the field. The papers on techniques only provide an introduction to the subject and an indication where details on methodology can be found. The case studies are taken from the files of working professional engineering geologists, or consultants in the subject who are now in academia, although many have a background of industrial experience. Most of the case studies have not been published before, those that have either been expanded to highlight the importance of land surface evaluation (e.g. **Hearn, Blong & Humphreys**; **Edmonds**) or they represent a key development in the subject (e.g. **Jones**).

The material presented in this report confirms that Land Surface Evaluation is established as part of essential good practice in land care, management and the development of the earth's surface. Since the first report (Anon 1982), it has become clear that the most important achievement is the realization that land surface data can be easily integrated with all other ground information. The culmination of this is the production of a geological ground model that embraces the work of all earth science disciplines (Fookes 1997). In the case studies it is apparent that the over-riding objective, and the main purpose of employing land surface evaluation techniques, is the development of an accurate ground model that minimizes the unknowns, identifies potential and actual problems, avoids hazards and maximizes the opportunities provided by a site. Carrying out an effective land surface evaluation for engineering projects prior to going on to site allows planning decisions to be made, ground investigations to be designed in the most cost-effective way, and preliminary designs to be drafted. The ground model developed provides a framework within which all subsequent data can be checked and analysed, as well as allowing extrapolations between data points to be made

based on a sound scientific rationale. In this way the local characteristics can be set within a general synoptic appreciation of the wider situation and factors that might influence a possible development site from outside the immediate area can be taken into account.

The First Working Party envisaged land surface evaluation being primarily the combination of geomorphological mapping and aerial photograph interpretation (i.e. remote sensing). This would be carried out based on a framework of land classification and supported by comprehensive desk studies. Comparing the examples presented in the two working party reports, the importance of these techniques has continued. These techniques are exemplified in the second report by a series of case studies:

- **Shilston, Harrison & French** (desk studies for the Docklands light railway);
- **Hearn** (aerial photograph interpretation for road construction in mountainous terrain);
- **Birch** (geomorphological mapping for gas pipelines in South Wales);
- **Phipps** (land classification for the Channel Tunnel high speed rail link);
- **Fookes, Lee & Sweeney** (terrain evaluation for a pipeline in Algeria).

However, since the first report there has been a vast increase in the techniques associated with remote sensing which are now beginning to impinge on engineering projects. Whilst satellite image interpretation has only really found a place on very large regional projects, aerial photography and spatial data handling systems have undergone significant changes. Of particular interest are the developments in digital photogrammetry (**Chandler**), and Geographical Information Systems (GIS) (**Nathanail & Symonds**; **Hearn, Hodgson & Woddy**). However, over the next decade we expect to see the full application of stereo-satellite imagery, pre-defined multi-spectral waveband sensing on aerial surveys, synthetic aperture radar, and airborne laser scanning. In addition, with the advent of more work in the nearshore and offshore zones, the present standard techniques of seabed remote sensing such as seismic profiling, bathymetric and side-scan sonar survey will be increasingly utilized to establish geological ground models.

From: GRIFFITHS, J. S. (ed.) *Land Surface Evaluation for Engineering Practice*. Geological Society, London, Engineering Geology Special Publications, **18**, 241–243. 0267-9914/01/$15.00 © The Geological Society of London 2001.

The other area where the case studies in the second report clearly show development since the first report, is in the growing recognition that 'geomorphology', as one of the critical disciplines within land surface evaluation studies, is not merely the science of 'morphology'. It is now established that geomorphology provides a basis for understanding the processes that create and maintain 'landforms.' Unfortunately, for many engineers the only useful product of a geomorphological survey was a map, and in order to understand the dynamics of the landscape they would turn to other disciplines such as soil mechanics, hydraulic engineering and hydrology. However, the case studies by **Lee & Brunsden** (coastal sediment dynamics) and **Edmonds** (subsidence in Chalk) demonstrate that the broader view that geomorphologists have about landscape development has a significant role to play in engineering projects. This broader geomorphological view takes into account space and time, inheritance of long-term trends, evolution of landscape, extreme events, and environmental change. These are factors that are only now beginning to be recognized as critical in the proper investigation of terrain in an engineering context. It's for this reason that the second working party envisages a much closer integration of engineering geology and geomorphology in the future, as demonstrated by the case studies of the Channel Tunnel (**Griffiths**), the Cardiff Barrage (**Edwards**) and the 'Ave' railway study in Spain (**Birch**).

The major change that has taken place between the first and second working party reports, and one that was expected, has been in the development of new techniques in land surface evaluation. These now play a role in engineering at all levels from planning, through ground investigation, design, and construction to remediation.

The first identifiable change is that land surface or geomorphological evaluations are now being used in actual construction work. **Martin** exemplifies this on the design of remedial works for the Dharan–Dhankuta road in Nepal, and **Hearn** in the general construction and rehabilitation of low-costs roads in mountainous areas. The development of techniques to evaluate the performance of structures in the natural environment (**Charman, Carey & Fookes**) is a further refinement of this process. These examples demonstrate how the civil works can only be effective if carried out with a full understanding of the natural earth surface processes that are active in the contemporary landscape.

The next major development has been in the field of hazard studies. Anticipating future geomorphological events, particularly extreme events with a very significant environmental impact, has come to be a major part of land surface evaluation practice. These hazard studies have dealt with a range of problems including mass movement (**Hearn & Griffiths; Charman; Petley; Shilston**), flood flows and debris flow runout (**Hearn, Blong & Humphreys**), ground subsidence (**Edmonds; Lee & Sakalas**), and sand and dust (**Jones**). It is anticipated that hazard studies and, increasingly, risk management will continue to grow in importance for practitioners in land surface evaluation at all stages of future and existing engineering development projects.

Another new technique that has been developed is ground behaviour analysis. This involves mapping landform types that are classified by genesis or mechanism, and plotting the known vectors or rates of displacement. In the report, the case study presented by **Lee & Moore** demonstrates the application of this powerful new technique that clearly has many potential applications in planning and engineering.

In addition to these broad themes, there is a wide range of applications demonstrated by the case studies including resource survey (**Allison**), defining the borders of countries in military conflict (**Nathanail**), and establishing geological ground models for tunnels (**Fookes & Shilston**). Whilst the case studies presented in the second working party report do not cover a full range of climatic regions, examples in the literature demonstrate the applicability of the land surface evaluation techniques to all types of environmental conditions. For example, glacial hazards are not dealt with by the case studies in this report but Reynolds (1998) examines aspects of these problems. Clearly the range and type of applications will continue to expand, although it should be recognized that many of the best examples of good practice will remain in consultants files. Finding time and getting permission to publish such work will continue to be difficult.

The future

In view of the proven success of land surface evaluation over thirty years, it remains a disappointment that many engineering failures still occur because the nature of the natural terrain has not been understood (Hutchinson 2001). Land surface evaluation is an integral component in the development of the Fookes (1997) geological ground model and as such is a key aspect of site investigation for all engineering works. The omission of any reference to land surface evaluation, and indeed the lack of detailed reference to any form of ground mapping, in the latest issue of the Code of Practice for Site Investigation BS 5930:1999 (BSI 1999) is worrying and should be rectified.

Ground investigations are expensive and systematic land surface evaluations remain one of the best ways of ensuring clients get value for their money. Professor P G Fookes in his 1997 Glossop lecture provided the format for carrying out site investigations (see Fookes 1997, figure 42). This format contained one task entitled 'Site Reconnaissance/and Survey' that, along with local knowledge and experience, fed into the preparation of preliminary geological/geomorphological maps/sections. Land surface evaluation has to be regarded as critical to this task, but it should also be emphasized that a land surface evaluation perspective needs to be included in all phases of a development. In order to achieve this practitioners are advised to review the techniques and case studies included in this report as they provide examples both of the tools that are available and the way they have been successfully applied to engineering projects over the past few decades.

As a final statement, however, the Working Party came to one very clear conclusion. The best way to understand the problems of the ground is still to walk over it, to learn to observe, to record the observations and measurements, to formulate questions and hypotheses, to design investigations, and to discuss the results in an informed way using a good scientific rationale. This is the message of this book.

References

ANON. 1982. Land surface evaluation for engineering practice. *Quarterly Journal of Engineering Geology*, **15**, 265–316.

BRITISH STANDARDS INSTITUTION. 1999. *Code of Practice for Site Investigations.* BS 5930:1999, British Standards Institution, London.

FOOKES, P. G. 1997. Geology for engineers: the geological model, prediction, and performance. *Quarterly Journal of Engineering Geology*, **30**, 293–424.

HUTCHINSON, J. N. 2001. Reading the ground: morphology and geology in site appraisal. Fourth Glossop Lecture. *Quarterly Journal of Engineering Geology and Hydrogeology*, **34**.

REYNOLDS, J. M. 1998. High-altitude glacial lake hazard assessment and mitigation: a Himalayan perspective. *In*: MAUND, J. G. & EDDLESTON, M. (eds) *Geohazards in Engineering Geology.* Geological Society, Engineering Geology Special Publications, **15**, 25–34.

Index